Shape memory materials are fascinating materials, with potential for application as 'smart materials' and also as new functional materials. This book presents a systematic and up-to-date account of all aspects of shape memory materials, from fundamentals to applications.

Starting from the basic principles of the martensitic transformation, on which the shape memory effect and the superelasticity of alloys are based, the mechanisms of the two phenomena are clearly described, together with possible applications. The characteristics, fabrication techniques and thermo-mechanical treatment of various shape memory alloys are described in detail, with special emphasis on Ti–Ni and Ti–Ni–X (with X being Cu, Fe, etc.) alloys. The book then describes various applications and design principles, for example in actuators, medical applications and as smart materials. The book contains chapters on shape memory ceramics and polymers as well as shape memory alloys, making the book a comprehensive account of the field.

This volume will be of interest to researchers and graduate students in materials science and engineering and mechanical engineering with an interest in smart or intelligent materials.

SHAPE MEMORY MATERIALS

SHAPE MEMORY MATERIALS

Edited by

K. OTSUKA

Institute of Materials Science, University of Tsukuba

and

C. M. WAYMAN

*Department of Materials Science and Engineering,
University of Illinois*

CAMBRIDGE
UNIVERSITY PRESS

PUBLISHED BY THE PRESS SYNDICATE OF THE UNIVERSITY OF CAMBRIDGE
The Pitt Building, Trumpington Street, Cambridge, United Kingdom

CAMBRIDGE UNIVERSITY PRESS
The Edinburgh Building, Cambridge CB2 2RU, UK   www.cup.cam.ac.uk
40 West 20th Street, New York, NY 10011-4211, USA   www.cup.org
10 Stamford Road, Oakleigh, Melbourne 3166, Australia
Ruiz de Alarcón 13, 28014, Madrid, Spain

First published 1998
First paperback edition (with corrections) 1999

Typeset in 11/14 pt Times New Roman [vN]

A catalogue record for this book is available from the British Library

Library of Congress Cataloguing in Publication data

Shape memory materials / edited by K. Otsuka and C. M. Wayman
p..   cm.
ISBN 0 521 44487 X (hc)
1. Shape memory alloys.   I. Otsuka, Kazuhiro, 1937–
II. Wayman, Clarence Marvin, 1930–
TA487.548   1998
620.1′632–dc21   97–38119   CIP

ISBN 0 521 44487 X hardback
ISBN 0 521 663849 paperback

Transferred to digital printing 2002

Contents

vii

Contributors

Masahiro IRIE, Professor
Department of Chemistry and Biochemistry, Graduate School of Engineering, Kyushu University, 6-10-1, Hakozaki, Higashi-Ku, Fukuoka 812-8581, Japan

Tadashi MAKI, Professor
Department of Materials Science and Engineering, Kyoto University, Yoshida-hom-machi, Sakyo-ku, Kyoto 606-8501, Japan

Keith MELTON, Research Director
British Steel plc, Swinden Technology Centre, Moorgate, Rotherham, South Yorkshire, S60 3AR, UK

Shuichi MIYAZAKI, Professor
Institute of Materials Science, University of Tsukuba, Tsukuba, Ibaraki 305-8573, Japan

Ichizo OHKATA, General Manager
Medical Devices Division, PIOLAX INC. 179, Kariba-cho, Hodogaya-ku, Yokohama 240-0025, Japan

Kazuhiro OTSUKA, Professor
Institute of Materials Science, University of Tsukuba, Tsukuba, Ibaraki 305, Japan

Toshio SABURI, Professor
Department of Materials Science and Engineering, Faculty of Engineering, Osaka University, 2-1, Yamadaoka, Suita, Osaka 565-0871, Japan

Rudi STALMANS, Post-doctoral Fellow
Departement Metaalkunde en Toegepaste Materiaalkunde, Katholieke Universiteit Leuven, W. de Croylaan 2, B-3001 Heverlee, Belgium

Yuichi SUZUKI, General Manager
Yokohama R&D Laboratories, The Furukawa Electric Co. Ltd., 2-4-3, Okano, Nishi-ku Yokohama 220-0073, Japan

Tsugio TADAKI, Professor
Department of Environmental Science, Faculty of Science, Osaka Women's University, 2-1 Daisen-cho, Sakai, Osaka 590-0035, Japan

Kenji UCHINO, Professor
*Director, International Center for Actuators and Transducers, 134 Materials Research
Laboratory, The Pennsylvania State University, University Park, PA 16802, USA*

Jan VAN HUMBEECK, Associate Professor
*Departement Metaalkunde en Toegepaste Materiaalkunde, Katholieke Universiteit
Leuven, W. de Croylaan 2, B-3001 Heverlee, Belgium*

C. Marvin WAYMAN, Professor Emeritus
*Department of Materials Science and Engineering, University of Illinois, Urbana, Illinois
61801, USA*

Preface

The shape memory alloys are quite fascinating materials characterized by a shape memory effect and superelasticity, which ordinary metals and alloys do not have. This unique behavior was first found in a Au–47.5 at% Cd alloy in 1951, and was publicized by its discovery in a Ti–Ni alloy in 1963. After much research and development thereafter, shape memory alloys are now being practically used as new functional alloys for pipe couplings, antennae for cellular phones and various actuators in electric appliances, etc. Furthermore, they have attracted keen attention as promising candidates for smart materials, since they function as sensors as well as actuators. Under these circumstances, many international conferences and symposia have been held on the subject, and as a result many proceedings have been published. Despite this fact, systemically written books on the subject are very few, especially in English. This was the main motivation for the editors' plan to publish the present book. The purpose of this book is to provide details of all aspects of the shape memory alloys from fundamentals to applications on an intermediate level.

Since the shape memory effect and superelasticity are based on the diffusionless phase transformation called the martensitic transformation, the basic notion and the characteristics of the transformation, along with the mechanisms of the shape memory effect and superelasticity, are explained in the first two chapters. Then, the descriptions of various shape memory alloys follow with special emphasis on Ti–Ni alloys, which are the most important practical shape memory alloys. In the chapters on applications, the design principle for shape memory alloys is described in detail from a mechanistic point of view, along with fabrication techniques and characteristics of the alloys.

As another special feature of the present book, chapters on shape memory ceramics and polymers are included. The shape memory effect in the former is due to either paraelectric to ferroelectric or anti-ferroelectric to ferroelectric

xiii

transitions and is associated with much smaller shape memory strains compared to those of alloys, but the response is much faster than the latter, since the former is electrically driven. The shape memory effect in some polymers is due to the glass transition. Although polymers are not strong in the high temperature phase (rubber-state), they are non-electro-conductive. Thus, they are complementary to shape memory alloys in actual applications, although their origins and characteristics are quite different.

In the writing and editing of the present book, we tried to make it concise, clear and informative and especially up-to-date. We hope it will be useful for researchers both in fundamental and applied fields and graduate students in materials science and mechanical engineering courses, and will be useful for developing the field further.

Finally, we express our hearty thanks to Publishing Director, Dr Simon Capelin, for his interest, encouragement and patience, without which the present book would not have been produced.

K. Otsuka and C. M. Wayman

1

Introduction

K. OTSUKA AND C. M. WAYMAN

Symbols for Chapters 1 and 2

A_f reverse transformation finish temperature
A_s reverse transformation start temperature
B lattice deformation matrix
d_1 shape strain direction (unit column vector)
F_d diagonal matrix
F_s symmetric matrix
g^m chemical free energy of the martensitic phase per unit volume
g^p chemical free energy of the parent phase per unit volume
$\Delta g_c = g^m - g^p$ chemical free energy change upon martensitic transformation per unit volume
G^m Gibbs free energy of the martensitic phase
G^p Gibbs free energy of the parent phase
$\Delta G^{p-m} = G^m - G^p$
ΔG^s free energy change due to an applied stress
$\Delta G_c = \Delta G^{p-m}$ chemical free energy change upon martensitic transformation
ΔG_e strain energy arising from martensitic transformation
$\Delta G_{nc} = \Delta G_s + \Delta G_e$ non-chemical term in the free energy change upon martensitic transformation
ΔG_s surface energy arising from martensitic transformation
H hexagonal (Ramsdel notation)
ΔH^* enthalpy of transformation
I identity matrix
K_1 twinning plane
K_2 undistorted plane
m_1 magnitude of shape strain
m_1^n dilatational component of shape strain ($\Delta V/V$)
m_1^b shear component of shape strain

M_f martensite finish temperature
M_s martensite start temperature
p hydrostatic pressure
p_1' invariant plane normal (habit plane normal) (p_1' unit row vector)
P_1 shape strain matrix
P_2 lattice invariant shear matrix
R rhombohedral (Ramsdel notation)
R rotation matrix
s twinning shear
ΔS entropy of transformation
T temperature
T_0 equilibrium temperature between parent and martensite
V volume of the parent phase
ΔV volume change upon martensitic transformation
ε strain associated with the martensitic transformation
ε_c calculated transformation strain
η_1 twinning shear direction
η_2 the intersection of the plane of shear and the K_2 plane
λ an angle between the shear component of the shape strain and the tensile axis
σ applied stress
σ_n normal component of an applied stress along the habit plane normal
τ shear stress component of an applied stress along the shear component of the shape strain
χ an angle between the habit plane and the tensile axis

1.1 Invitation to shape memory effect and the notion of martensitic transformation

The shape memory effect (to be abbreviated SME hereafter) is a unique property of certain alloys exhibiting martensitic transformations, as typically

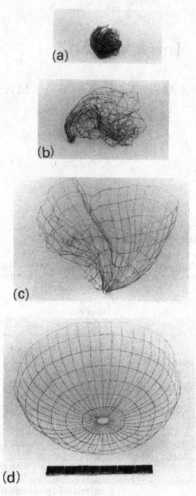

Fig. 1.1. Demonstration of shape memory effect by a space antenna of Ti–Ni wires.
The antenna deformed in the martensitic state in (a) reverts to the original shape
(b–d) upon heating by solar heat. (Courtesy Goodyear Aerospace Corporation)

shown in Fig. 1.1. Even though the alloy is deformed in the low temperature
phase, it recovers its original shape by the reverse transformation upon heating
to a critical temperature called the reverse transformation temperature. This
effect was first found in a Au–47.5 at% Cd alloy by Chang and Read,[1] and then
it was publicized with the discovery in Ti–Ni alloys by Buehler *et al.*[2] Many
other alloys such as In–Tl,[3,4] Cu–Zn[5] and Cu–Al–Ni[6] were found between the
above two and thereafter. (See Ref. [7] for historical developments.) The same
alloys have another unique property called 'superelasticity' (SE) at a higher
temperature, which is associated with a large (several–18%) nonlinear recoverable strain upon loading and unloading. Since these alloys have a unique

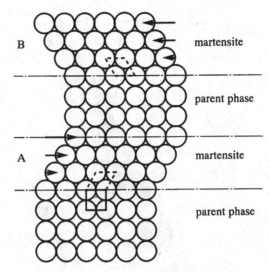

Fig. 1.2. A simplified model of martensitic transformation. See text for details.

property in remembering the original shape, having an actuator function and having superelasticity, they are now being used for various applications such as pipe couplings, various actuators in electric appliances, automobile applications, antennae for cellular phones and medical implants and guidewires etc. Besides, since they have the function of an actuator as well as a sensor, they are promising candidates for miniaturization of actuators such as microactuators or micromachines or robots. These will be discussed in detail in later chapters.

Since both SME and SE are closely related to the martensitic transformation (MT), the basic notion of MT is first given in a very naive and over-simplified manner. More accurate descriptions will follow in the next section. The martensitic transformation is a diffusionless phase transformation in solids, in which atoms move cooperatively, and often by a shear-like mechanism. Usually the parent phase (a high temperature phase) is cubic, and the martensite (a lower temperature phase) has a lower symmetry. The transformation is schematically shown in Fig. 1.2. When the temperature is lowered below some critical one, MT starts by a shear-like mechanism, as shown in the figure. The martensites in region A and in region B have the same structure, but the orientations are different. These are called the correspondence variants of the martensites. Since the martensite has a lower symmetry, many variants can be formed from the same parent phase. Now, if the temperature is raised and the martensite becomes unstable, the reverse transformation (RT) occurs, and if it is crystallographically reversible, the martensite reverts to the parent phase in the original orientation. This is the origin of SME, which will be described in

Fig. 1.3. Optical micrograph of spear-like γ_1' martensite in Cu–14.2 mass%
Al–4.2 mass% Ni alloy.

more detail later. The above example clearly shows that the characteristics of
MT lie in the cooperative movement of atoms. Because of this nature, MT is
sometimes called the displacive transformation or military transformation,
which are equivalent in usage to MT. Thus, even though the relative atomic
displacements are small (compared with inter-atomic distance), a macroscopic
shape change appears associated with MT, as seen in Fig. 1.2. It is in this
respect that MT is closely related to SME and SE.

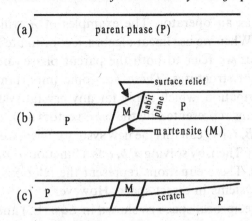

Fig. 1.4. Schematic representation of shape change associated with martensitic transformation; (a) original parent single crystal, (b) surface relief due to transformation, (c) change in direction of pre-scratched straight line upon martensitic transformation.

Figure 1.3 shows an optical micrograph of a typical martensite in a Cu–Al–Ni alloy. The flat region in light contrast represents a parent phase, while the plate-like morphology of martensite variants is observable by surface relief (surface upheaval) effects. The thin band contrasts in each martensite variant are twins, which will be discussed later. Similarly, when a straight line is marked on the surface of a specimen, the line changes direction upon MT, as schematically shown in Fig. 1.4(c). These experiments clearly show that the shape change associated with MT is linear, since upon MT, a line and a surface are changed into another line and surface, respectively. This means that the shape change associated with MT can be described by a matrix *as an operator*.

1.2 Martensitic transformations: crystallography

1.2.1 Linear algebra describing a deformation (mathematical preparation)

When a deformation is linear, the deformation is represented by the following equation.

$$\begin{pmatrix} x_2 \\ y_2 \\ z_2 \end{pmatrix} = \begin{pmatrix} a_{11} & a_{12} & a_{13} \\ a_{21} & a_{22} & a_{23} \\ a_{31} & a_{32} & a_{33} \end{pmatrix} \begin{pmatrix} x_1 \\ y_1 \\ z_1 \end{pmatrix}. \tag{1.1}$$

In a compact form, it may be written as

$$r_2 = Ar_1, \tag{1.2}$$

where A represents the matrix a_{ij} etc. That is, any vector r_1 is transformed into

r_2 by a matrix A as an operator. The examples of A will be given in the following sections. When we use linear algebra, a coordinate transformation is also necessary, since we refer to both the parent phase and the martensite, which have different structures. There are some important formulas for a coordinate transformation, which apply for any crystal system. Suppose we have two axis systems represented by the base vectors a, b, c (we call this the old system) and A, B, C (we call this the new system). Then, we can immediately write down Eq. (1.3). Then, by solving a, b, c as a function of A, B, C in Eq. (1.3), we obtain Eq. (1.4). These equations represent the relations between the old and the new axis systems in direct space. However, it is proven that similar equations hold in reciprocal space as shown in Eqs. (1.5) and (1.6), where a^*, b^*, c^* represent the base vectors in reciprocal space, which correspond to a, b, c, and the same thing applies to A^*, B^*, C^*.[8]

$$
\begin{aligned}
A &= s_{11}a + s_{12}b + s_{13}c \\
B &= s_{21}a + s_{22}b + s_{23}c \\
C &= s_{31}a + s_{32}b + s_{33}c
\end{aligned}
\tag{1.3}
$$

$$
\begin{aligned}
a &= t_{11}A + t_{12}B + t_{13}C \\
b &= t_{21}A + t_{22}B + t_{23}C \\
c &= t_{31}A + t_{32}B + t_{33}C
\end{aligned}
\tag{1.4}
$$

$$
\begin{aligned}
A^* &= t_{11}a^* + t_{21}b^* + t_{31}c^* \\
B^* &= t_{12}a^* + t_{22}b^* + t_{32}c^* \\
C^* &= t_{13}a^* + t_{23}b^* + t_{33}c^*
\end{aligned}
\tag{1.5}
$$

$$
\begin{aligned}
a^* &= s_{11}A^* + s_{21}B^* + s_{31}C^* \\
b^* &= s_{12}A^* + s_{22}B^* + s_{32}C^* \\
c^* &= s_{13}A^* + s_{23}B^* + s_{33}C^*
\end{aligned}
\tag{1.6}
$$

Furthermore, there are other important formulas for arbitrary vector components in direct and in reciprocal space, as shown below.

$$
\begin{pmatrix} x \\ y \\ z \end{pmatrix} =
\begin{pmatrix} s_{11} & s_{21} & s_{31} \\ s_{12} & s_{22} & s_{32} \\ s_{13} & s_{23} & s_{33} \end{pmatrix}
\begin{pmatrix} X \\ Y \\ Z \end{pmatrix}
\tag{1.7}
$$

old A B C new

$$\begin{pmatrix} H \\ K \\ L \end{pmatrix} = \begin{pmatrix} s_{11} & s_{12} & s_{13} \\ s_{21} & s_{22} & s_{23} \\ s_{31} & s_{32} & s_{33} \end{pmatrix} \begin{pmatrix} h \\ k \\ l \end{pmatrix} \qquad (1.8)$$

new   old

where xyz and XYZ refer to the direct lattice (i.e. direction) and hkl and HKL to the reciprocal lattice (i.e. plane).

When a coordinate transformation is applied for a vector or a plane, an operator is also transformed following the similarity transformation given below. (The proof is simple. See Ref. [9] for the proof.)

$$\bar{A} = R^{-1}AR \text{ or } A = R\bar{A}R^{-1} \qquad (1.9)$$
$$\bar{A} = R^{T}AR \text{ or } A = R\bar{A}R^{T} \text{ (when } R \text{ is orthogonal),} \qquad (1.10)$$

where A is an operator in the old system, while $\bar{A}$ is that in the new system, R is a rotation matrix, and R^{-1} is an inverse of R, and R^{T} is R transposed. In the martensite crystallography calculations, the similarity transformation is often used, since operators often refer to the parent phase in such calculations.

1.2.2 Structural change without diffusion: lattice correspondence, correspondence variant and lattice deformation

We now discuss how the martensite crystal is produced from the parent crystal without diffusion. As a typical example, we describe this for the well-known FCC (Face-Centered Cubic)–BCT (Body-Centered Tetragonal) transformation in steels. Figure 1.5(a) shows two FCC unit cells, in which we notice a BCT lattice with the axial ratio $c/a = \sqrt{2}$. Thus, if the X and Y axes in the figure are elongated and the Z axis is contracted so that c/a becomes the value of the martensite (i.e. a value close to 1), then a BCT martensite is created, as shown in Fig. 1.5(b). This is the mechanism originally proposed by Bain.[10] Although the mechanism is different from one alloy to another, it is always possible to create a martensite from a parent by the combination of elongation, contraction and shear along certain directions. If the lattice parameter of the FCC is a_0, and those of the BCT are a and c, then the lattice deformation matrix with respect to the XYZ axes is written as follows.

$$B = \begin{pmatrix} \sqrt{2}a/a_0 & 0 & 0 \\ 0 & \sqrt{2}a/a_0 & 0 \\ 0 & 0 & c/a_0 \end{pmatrix} \qquad (1.11)$$

By utilizing the similarity transformation, the lattice deformation matrix with

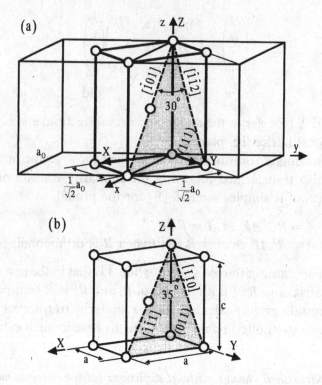

Fig. 1.5. Mechanism of FCC–BCT (or BCC) transformation by Bain. *xyz* represents
the crystal axes in the parent FCC lattice, while *XYZ* represents those axes in the
BCT martensite. See text for details.

respect to the parent lattice (i.e. the *xyz* axes in (a)) is given as follows.

$$B = R\bar{B}R^{\mathrm{T}}$$

$$= \begin{pmatrix} 1/\sqrt{2} & 1/\sqrt{2} & 0 \\ -1/\sqrt{2} & 1/\sqrt{2} & 0 \\ 0 & 0 & 1 \end{pmatrix} \begin{pmatrix} \sqrt{2}a/a_0 & 0 & 0 \\ 0 & \sqrt{2}a/a_0 & 0 \\ 0 & 0 & c/a_0 \end{pmatrix} \begin{pmatrix} 1/\sqrt{2} & -1/\sqrt{2} & 0 \\ 1/\sqrt{2} & 1/\sqrt{2} & 0 \\ 0 & 0 & 1 \end{pmatrix}.$$

$$(1.12)$$

Another important notion is the lattice correspondence, which is associated
with the lattice deformation. Since MT is diffusionless, there is a one-to-one
correspondence in the directions and planes between the parent and the
martensite. It is clear in the figure that $[1/2\ \overline{1/2}\ 0]_{\mathrm{p}}$ and $[1/2\ 1/2\ 0]_{\mathrm{p}}$ correspond
to $[100]_{\mathrm{m}}$ and $[010]_{\mathrm{m}}$, respectively, where the subscripts p and m correspond
to parent and martensite respectively. Then, how an arbitrary $[xyz]_{\mathrm{p}}$ and $(hkl)_{\mathrm{p}}$
in the parent phase correspond to which $[XYZ]_{\mathrm{m}}$ and $(HKL)_{\mathrm{m}}$ in the martensite is the main problem here. This is essentially the problem of coordinate

transformation. In Fig. 1.5(a), it is easy to write down the coordinate transformation between the xyz system and the XYZ system. When we transform from the XYZ system in (a) to the XYZ system in (b), lattice change (by lattice deformation) occurs, but the Miller indices are invariant under this transformation. Thus, the above obtained equation of coordinate transformation still holds, as follows.

$$\begin{pmatrix} x \\ y \\ z \end{pmatrix} = \begin{pmatrix} 1/2 & 1/2 & 0 \\ -1/2 & 1/2 & 0 \\ 0 & 0 & 1 \end{pmatrix} \begin{pmatrix} X \\ Y \\ Z \end{pmatrix}, \qquad \begin{pmatrix} H \\ K \\ L \end{pmatrix} = \begin{pmatrix} 1/2 & -1/2 & 0 \\ 1/2 & 1/2 & 0 \\ 0 & 0 & 1 \end{pmatrix} \begin{pmatrix} h \\ k \\ l \end{pmatrix}. \quad (1.13)$$

old new new old

From these equations, we can easily find that $[\bar{1}01]_p$ corresponds to $[\bar{1}\bar{1}1]_m$, $[\bar{1}\bar{1}2]_p$ to $[0\bar{1}1]_m$ and $(111)_p$ to $(011)_m$, by lattice correspondence.

One more important notion associated with lattice correspondence is the correspondence variant (c.v.). In Fig. 1.5, we chose the z axis as the c axis of the martensite. We could equally choose the x and y axes as the c axis of the martensite. Thus, three correspondence variants are possible in the FCC–BCT transformation.

Among many structural changes in various MTs, those in β-phase alloys are important. Thus, they are briefly described below. The β-phase alloys, such as Au–Cd, Ag–Cd, Cu–Al–(Ni), Cu–Zn–(Al) etc., are characterized by the value of electron/atom ratio $e/a \approx 1.5$, at which BCC (Body-Centered Cubic) or ordered BCC structure is stabilized due to nesting at the Brillouin zone boundary. The ordered BCC structures are usually B2 type or D0$_3$ type. With lowering temperature, these ordered BCC structures change martensitically into close-packed structures, which are called 'long period stacking order structures' with a two-dimensional close-packed plane (basal plane), since the entropy term in the Gibbs free energy becomes negligible at low temperatures and the decrease of internal energy becomes more important. Following Nishiyama and Kajiwara,[11] we describe the structural change for the B2 type parent phase using Fig. 1.6, but that for the D0$_3$ type parent phase is similar. The structure of the B2 type parent phase is shown in (a). This structure may be viewed as that in which the $(110)_{B2}$ plane is stacked in $A_1B_1A_1B_1 \ldots$ order, as shown in (b). Upon MT, the $(110)_{B2}$ plane changes into a more close-packed plane $(001)_m$ in (c), by contracting along $[001]_{B2}$ and elongating along $[\bar{1}10]_{B2}$, so that the indicated angle changes from $70° 32'$ to close to $60°$. Once the plane becomes a close-packed one as shown in (c), we have three stacking positions A, B, C shown in (c). Then, we can create various stacking order structures. Theoretically, we can create an infinite number of long period stacking order

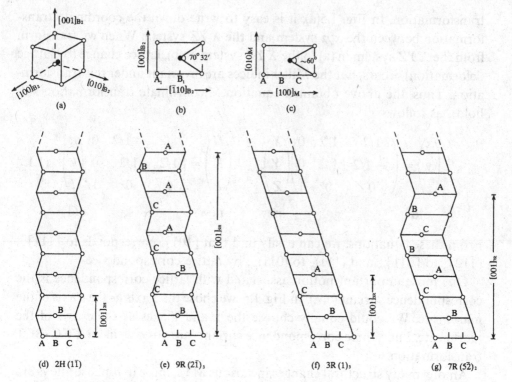

Fig. 1.6. The structural change from B2 parent phase into various martensites with long period stacking order structures. See text for details.

structures, but practically the first three (d–f) are most common, and the fourth (g) is a new one. There are two notations to describe these long period stacking order structures, i.e. Ramsdel notation and Zdanov symbol.[12-14] The meanings of these notations are as follows. In (d), the period along the c-axis is 2, and the symmetry is hexagonal, when ordering is disregarded. Thus, it is called 2H in Ramsdel notation. The Zdanov symbols in parentheses represent the number of layers in a clockwise sequence (positive number) and that in an anticlockwise sequence (negative number). Since the stacking sequence in the case of (d) is ABAB..., it is written as $(1\bar{1})$. In the case of (e), the period is 9, and the symmetry is rhombohedral. Thus, it is called 9R in Ramsdel notation. In the case of (g), 7R is the wrong notation, since it does not have rhombohedral symmetry, even though ordering is disregarded, but the term is commonly used erroneously. The usage of the two notations for (f) will be easily understood. In the above, the usage of Ramsdel notation is slightly different between Refs. [12,13] and Ref. [14]. We followed Refs. [12,13] by Kakinoki in the above. Thus care should be taken. There is another new notation proposed by the present authors[15] which is more logical and more accurate. However, we have omitted to describe it here, considering the character of this book and space limitation. For more details on MTs in β-phase alloys, see Refs. [16,17].

1.2.3 Lattice invariant shear and deformation twinning

Since MT is a first order transformation, it proceeds by nucleation and growth. Since MT is associated with a shape change as described above, a large strain arises around the martensite when it is formed in the parent phase. To reduce the strain is essentially important in the nucleation and growth processes of MT. There are two ways to attain it; either by introducing slip (b) or by introducing twins (c), as shown in Fig. 1.7(b) and (c). These are called the lattice invariant shear (LIS), since neither process changes the structure of the martensite. That is, either slip or twinning is a necessary process in MT for the above reason, and twins or dislocations are usually observed in martensites under electron microscopy. Which of slip or twinning is introduced depends upon the kind of alloys, but twinning is usually introduced as a LIS in SMAs. Thus, twinning is described in more detail in the following. Two twin crystals are generally related by a symmetry operation with respect to a mirror plane or a rotation axis. In deformation twinning, a twin is created by a proper shear, while twins are introduced upon MT for the above reason, and they can act as

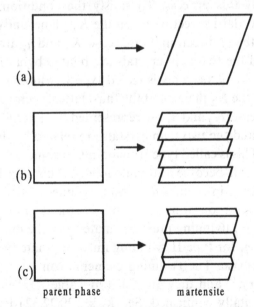

parent phase   martensite

Fig. 1.7. Schematically shows why the lattice invariant shear is required upon martensitic transformation; (a) shape change upon martensitic transformation; (b) and (c) represent the accommodation of strain by introducing slip (b) or twins (c), respectively.

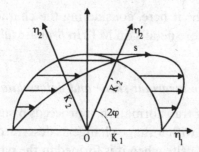

Fig. 1.8. The deformation of a unit sphere into an ellipsoid by shear, and the definition of K_1, K_2, η_1, η_2 and s. See text for details.

a deformation mode under stress. In this connection twins have a close relationship with SME. Now, when we discuss deformation twinning, we use a unit sphere and an ellipsoid as a result of a shear, as shown in Fig. 1.8. In this shearing process, K_1 and η_1 represent the shear plane and the direction of the shear, respectively. Obviously K_1 is an invariant plane and K_2 is another undistorted plane in this process. The plane which is normal to K_1 and is parallel to η_1 is called the plane of shear, and the intersection of K_2 and the plane of shear is called η_2. K_1, K_2, η_1, η_2 and a twinning shear s are called the twinning elements. In order to create a twin by this process, the original lattice must be restored by this process. To satisfy the condition, there are two cases.[18] In case I, two lattice vectors lie on the K_1 plane, and the third lattice vector is parallel to the η_2 direction. In this case, K_1 and η_2 are represented by rational indices, and the two twin crystals are related by a mirror symmetry with respect to the K_1 plane. This is called type I twinning. In case II, two lattice vectors lie on the K_2 plane, and the third lattice vector is parallel to the η_1 direction. In this case, K_2 and η_1 are represented by rational indices, K_1 and η_2 being irrational, and the two twin crystals are related by the rotation by π around the η_1 axis. This is called type II twinning. In some crystal systems, K_1, K_2, η_1 and η_2 may all become rational indices. This is called compound twinning, and the two twin crystals have both symmetry characteristics. With respect to the transformation twins as a lattice invariant strain, the following is proved:[19] K_1 for type I twinning must originate from the mirror plane in the parent phase, while η_1 for type II twinning must originate from the two-fold axis in the parent phase. The twinning elements can be calculated by the Bilby–Crocker theory,[20] and they are listed in Table 1.1[21] for various MTs, which are experimentally confirmed. See Refs. [20,22,32] for the details of deformation twinning.

Table 1.1. Twinning modes in martensite[21]

Structure	Twinning Elements					Alloys Observed
	K_1	K_2	η_1	η_2	s	
BCC (FCC → BCC)	$\{112\}$	$\{1\bar{1}2\}$	$\langle 1\bar{1}1\rangle$	$\langle 111\rangle$	$1/\sqrt{2}$	Fe–Ni, Fe–Pt
BCT (FCC → BCT)	$\{112\}$	$\{1\bar{1}2\}$	$\langle \bar{1}11\rangle$	$\langle 111\rangle$	$(2-\gamma^2)/\sqrt{2}\gamma$[a]	Fe–C, Fe–Ni–C, Fe–Cr–C
	$\{011\}$	$\{01\bar{1}\}$	$\langle 01\bar{1}\rangle$	$\langle 011\rangle$	$\gamma - 1/\gamma$	Fe–C
BCT (BCC → BCT)	$\{011\}$	$\{01\bar{1}\}$	$\langle 01\bar{1}\rangle$	$\langle 011\rangle$	$\gamma - 1/\gamma$	Au–Mn
FCT (FCC → FCT)	$\{011\}$	$\{01\bar{1}\}$	$\langle 01\bar{1}\rangle$	$\langle 011\rangle$	$\gamma - 1/\gamma$	In–Tl, Mn–Cu, In–Cd In–Pb, Fe–Pt, Fe–Pd
Tetragonal (B2 → 3R)	$\{111\}$	$\{1\bar{1}1\}$	$\langle 11\bar{2}\rangle$	$\langle 112\rangle$	$\sqrt{2}\gamma - 1/\sqrt{2}\gamma$	Ni–Al
HCP (BCC → HCP)	$\{10\bar{1}1\}$	irrational	irrational	$\{4\bar{1}\bar{5}3\}$	$\dfrac{\sqrt{4\gamma^4 - 17\gamma^2 + 27}}{2\sqrt{3}\gamma}$	Ti, Ti–Mo, Ti–V
Orthorhombic 2H (DO$_3$ → 2H)	$\{121\}$	$\{\bar{1},1.5036,0.5036\}$	$\langle \bar{1},0.7953,0.5907\rangle$	$\langle 111\rangle$	0.261[b]	Cu–Al–Ni, Cu–Al, Cu–Sn
	$\{\bar{1},1.5036,0.5036\}$	$\{121\}$	$\langle 111\rangle$	$\langle \bar{1},0.7953,0.5907\rangle$	0.261[b]	Cu–Al–Ni
2H (B2 → 2H)	$\{101\}$	$\{101\}$	$\langle 10\bar{1}\rangle$	$\langle 011\rangle$	0.0744[b]	Cu–Al–Ni
	$\{111\}$	$\{1,0.7073,1.2927\}$	$\langle 1,0.3740,0.6260\rangle$	$\langle 211\rangle$	0.156[c]	Au–Cd, Ag–Cd, Ti–Ta
Orthorhombic (BCC → Orthorhombic)	$\{\bar{1}11\}$	$\{0.522,0.823,0.221\}$	$\langle 0.788,0.579,0.209\rangle$	$\langle \bar{2}11\rangle$	0.19[d]	Ti–Ta
	$\{0.522,0.823,0.221\}$	$\{\bar{1}11\}$	$\langle \bar{2}11\rangle$	$\langle 0.788,0.579,0.209\rangle$	0.19[d]	
Monoclinic (B2 → Monoclinic)	$\{\bar{1}11\}$	$\{0.2470,0.5061,1\}$	$\langle 0.5404,0.4596,1\rangle$	$\langle \bar{2}11\rangle$	0.310[e]	Ti–Ni
	$\{\bar{1}11\}$	$\{0.6688,0.3375,1\}$	$\langle 1.5117,0.5117,1\rangle$	$\langle 211\rangle$	0.142[f]	Ti–Ni
	$\{0.7205,1,\bar{1}\}$	$\{011\}$	$\langle 011\rangle$	$\langle 1.5727,1,\bar{1}\rangle$	0.280[e]	Ti–Ni
	(001)	(100)	$[100]$	$[001]$	0.238	Ti–Ni

In the above table, irrational numbers depend upon the values of lattice parameters. Thus these values are shown for specific alloys indicated.

[a] $\gamma = c/a$ [b] Cu–Al–Ni [47]   [c] Au–Cd   [d] Ti–Ta [48]   [e] Ti–Ni [49]   [f] Ti–Ni [50]

1.2.4 Essence of the phenomenological theory of martensitic transformations

The final target of the theory of martensitic transformations from a crystallo-graphic point of view is to predict quantitatively all the crystallographic parameters associated with the transformations, such as habit plane and orientation relationship between parent and martensite etc. We now have two theories, the so-called 'phenomenological theories of martensitic transformations', which make the predictions possible. These were developed by Wechsler–Lieberman–Read (WLR)[24,25] and Bowles–Mackenzie (BM)[26] independently. Although the formulations are different, they are shown to be equivalent.[27] Since it is difficult to explain the theory in this short section, we describe only the basic ideas of the theory in order to understand the later sections, following the WLR theory, which is easier to follow physically.

The shape strain matrix P_1 as an operator, representing the whole MT is given as follows.

$$P_1 = \Phi_1 P_2 B, \tag{1.14}$$

where B represents the lattice deformation matrix to create a martensite lattice from a parent lattice, P_2 a lattice invariant shear matrix, and Φ_1 a lattice rotation matrix. In the theory, we focus attention on minimizing the strain energy associated with the transformation. Since such strains concentrate at the boundary between parent and martensite, we can eliminate such strains effectively by making the boundary an invariant plane, which is undistorted and unrotated on average. Since the invariant plane cannot be made only by B, P_2 is necessary, as schematically shown in Fig. 1.7. Although an undistorted plane can be made by the introduction of P_2, another Φ_1 is necessary to make it unrotated. Thus, P_1 is given by Eq. (1.14). That is, solving Eq. (1.14) under an invariant plane strain condition is the essence of the 'phenomenological theory', and the habit plane is given by the invariant plane. Although we omit the detailed descriptions, we continue to explain the basic ideas in the follow-ing. If we write $P_2 B = F$, it is always possible to separate F into the product of a symmetric matrix F_s and a rotation matrix Ψ, and the symmetric matrix can be diagonalized by the principal axis transformation. Thus,

$$F = \Psi F_s = \Psi \Gamma F_d \Gamma^{\mathrm{T}}, \tag{1.15}$$

where F_d is a diagonal matrix, Γ is a matrix for the diagonalization, and Γ^{T} is a transpose of Γ. By inserting this into Eq. (1.14), we obtain,

$$P_1 = \Phi_1 \Psi \Gamma F_d \Gamma^T. \tag{1.16}$$

In this expression, all matrices except for F_d are rotation matrices. Thus, only F_d carries the strain among them. Since F_d is a diagonal matrix, it is written as,

$$F_d = \begin{pmatrix} \lambda_1 & 0 & 0 \\ 0 & \lambda_2 & 0 \\ 0 & 0 & \lambda_3 \end{pmatrix}. \tag{1.17}$$

If we consider the distortion by F_d, it changes a unit sphere into an ellipsoid in the principal axis system, as shown in Fig. 1.9. Thus, the intersection between the sphere and the ellipsoid is not a plane in general. However, if and only if one of the following two conditions is satisfied, an undistorted plane exists.[28]

(1) One λ_i is 1, another is larger than 1, and the rest is smaller than 1 (e.g. $\lambda_1 < 1$, $\lambda_2 > 1$, $\lambda_3 = 1$).
(2) Two λ_i's are equal to 1.

Figure 1.9 shows the first case graphically, in which $\lambda_3 = 1$. We can recognize two undistorted planes. (We call the two solutions (+) and (−) solutions, following Saburi and Nenno[29] in the later section.) The condition that one λ_i is 1 requires that the twin width ratio $(1 - x):x$ or magnitude of slip as a LIS attains some specific value. That is, the invariant plane strain condition is satisfied only when the value of a LIS takes a specific value. Although we omit the details of the analysis, we can summarize the theory from another aspects. The input data necessary for the analysis are only the three following.

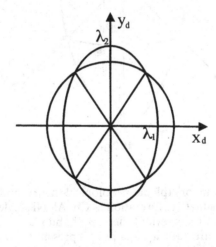

Fig. 1.9. The deformation of a unit sphere into an ellipsoid by F_d in the principal axis system. See text for details.

(1) The lattice parameters of parent and martensite.
(2) The lattice correspondence between parent and martensite.
(3) The lattice invariant shear.

By giving only three input data, we can calculate all the crystallographic parameters, such as habit plane, orientation relationship, shape strain, twin width ratio, orientation of K_1 plane, etc. Figure 1.10[30] shows an example of the critical comparison between theory and experiment for the $\beta_1(DO_3)-\gamma'_1$ (orthorhombic) transformation with type II twinning as a LIS in a Cu–Al–Ni single crystal. In the figure, closed symbols represent theoretical predictions and open ones experimental results. We observe excellent agreement between theory and experiment in all respects. See the figure caption for the details. In the past, the LISs were believed to be type I twinning, and there were several alloy systems in which theory and experiment did not agree well. However, since the LISs were found to be type II twinning in Cu–Al–Ni, Ti–Ni, Cu–Sn

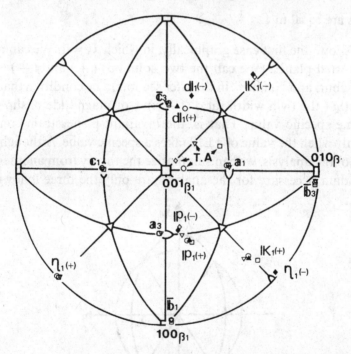

Fig. 1.10. Comparison of theory (phenomenological theory) and experiment for $\beta_1-\gamma'_1$ (orthorhombic) martensitic transformation in Cu–Al–Ni single crystals with $\langle 111 \rangle$ type II twinning as a lattice invariant shear; p_1: habit plane normal, K_1: K_1 plane normal, d_1: shape strain direction. a_1, b_1 and c_1 represent the orientation of c.v.1, and a_3, b_3 and c_3 that of c.v.3 relative to the parent phase. T.A. represents the tensile axis of the single crystals. The closed symbols represent theoretical results, the open symbols experimental ones. (After Okamoto *et al.*[30])

alloys, etc., the theory is now known to apply well, if the LIS is correctly chosen[31] (although there still remain some exceptional cases such as {225} transformation in an Fe–Cr–C alloy).

The martensitic transformation described by the phenomenological theory is that with an invariant plane, as described above. The general deformation with an invariant plane is an invariant plane strain, which consists of a shear component and a dilatational component, as shown in Fig. 1.11. The matrix representation of the invariant plane strain is obtained as follows. We represent the displacement at a unit distance from the habit plane m_1d_1 [d_1: a unit vector with a component $(d_1d_2d_3)$], and a unit vector normal to the habit plane p_1' ($p_1p_2p_3 = p_xp_yp_z$: a row vector). Then,

$$P_1x - x = m_1d_1(p_1'x)$$
$$P_1x = Ix + m_1d_1(p_1'x) = (I + m_1d_1p_1')x$$
$$\therefore P_1 = I + m_1d_1p_1' \tag{1.18}$$

where I is an identity matrix. By using the phenomenological theory, m_1, d_1, p_1' are all calculated. That is, upon MT, a complicated deformation, which consists of the lattice deformation, the lattice invariant shear and the lattice rotation, occurs following Eq. (1.14), but the end result is macroscopically represented by the invariant plane strain, as shown in Fig. 1.11.[32] Because of this, martensites usually appear as thin plates with an invariant plane as a habit plane. As shown in Fig. 1.11, the invariant plane strain can be resolved into two components, i.e. a shear component (m_1^p) and a dilataional component (m_1^n), which is equivalent to the volume change upon MT ($\Delta V/V$). By the way, in the thermoelastic transformation to be described in the next section, the

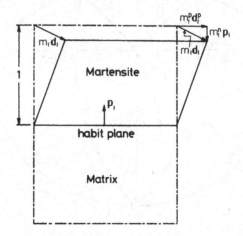

Fig. 1.11. Illustration of invariant strain, $P_1 = I + m_1d_1p_1'$ and its resolution into two components, $m_1^pd_1^p$ and $m_1^np_1'$. (After Otsuka and Wayman[46])

volume change is very small (e.g. $\sim 0.3\%$). Thus, the invariant plane strain is very close to a simple shear for the transformation. There are many good texts on martensitic transformations in general.[9,16,17,22] See Refs. [9,24–28] for the details of the phenomenological theory, and see Refs. [9,31] for the comparison between theory and experiment.

1.2.5 Self-accommodation of martensites

In the preceding section, we described the essence of the phenomenological theory. Following the theory, as many as 24 habit planes may be formed upon MT, even from a single crystal parent phase. We will describe this in the following for the B2–7R(14M) transformation in a Ni–37.0 at% Al alloy,[33] to be specific, where 14M is a new notation briefly mentioned in Section 1.2.2. This structural change was described in that section. Since there are six $\{110\}_p$ planes, which become the basal plane of the martensite, and there are two shearing directions, there are 12 correspondence variants. If we carry out the phenomenological theoretical calculations for the above 12 c.v., we obtain 24 habit planes, as shown in Fig. 1.12,[33] where the numbers label each c.v., and

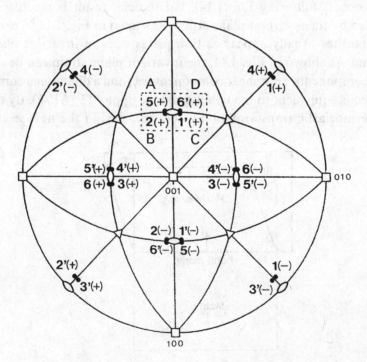

Fig. 1.12. 24 habit plane variants for B2–7R(14M) transformation in Ni–Al alloy calculated by phenomenological theory. See text for details. (After Murakami *et al.*[33])

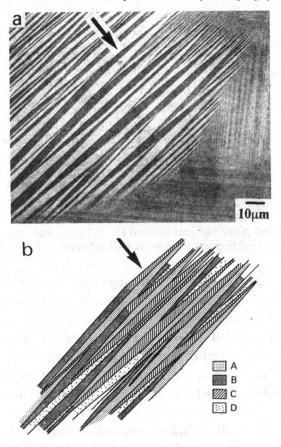

Fig. 1.13. (a) A typical SEM micrograph of self-accommodating 7R(14M) martensite. (b) Four habit plane variants (A, B, C, D) were identified by trace analysis. (After Murakami *et al.*[33])

(+) and (−) signs correspond to the two solutions described in the previous section, referring to Fig. 1.9. We call these habit plane variants (h.p.v.), distinct from c.v. Although we showed 24 h.p.v. for this particular case, 24 h.p.v. are quite common for the other types of transformations.[29]

We showed in the previous section that the invariant plane strain condition is very effective in reducing strains associated with the formation of a martensite plate. However, a shear component is still not eliminated under this condition, as is clear by checking Fig. 1.11. Thus, the second step strain accommodation is necessary, and this may become possible by the combination of favorable two or four h.p.v. This is called the self-accommodation of martensites. A typical SEM (scanning electron microscopy) micrograph is shown in Fig. 1.13(a),[33] in which four h.p.v. were confirmed to be present by trace analysis, as shown in Fig. 1.13(b), in which four h.p.v. designated as A, B,

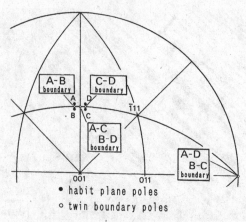

- habit plane poles
- twin boundary poles

Fig. 1.14. Four habit plane variants around the $\bar{1}01$ pole, and three types of twin boundaries. See text for details. (After Murakami *et al.*[33])

C, D were confirmed to correspond to the four h.p.v. around the $(\bar{1}01)_p$ pole as shown in Fig. 1.14, which is an enlarged portion of the square area in Fig. 1.12. Furthermore, it is confirmed that these h.p.v. are twin-related to each other,[29,33–36] i.e. A–C and B–D are type I twin-related, A–B and C–D are type II twin-related, and A–D and B–C are compound twin-related. That is, these h.p.v. are twins to each other, and their boundaries are mobile under stress, and thus the twinning or detwinning works as deformation modes in the martensitic state, as will be shown in the next chapter.

We now show that the above self-accommodation is very effective in reducing strains upon MT, by using the shape strain matrix P_1 as a measure of strains introduced by MT.[29] If P_1 is an identity matrix, the MT is strain-free. If a martensite plate of h.p.v.$1'(+)$ is produced, the result of the calculation of the shape strain matrix P_1 for the variant is,

$$P_1\{1'(+)\} = \begin{pmatrix} 0.946\,21 & 0.006\,33 & 0.055\,76 \\ -0.005\,69 & 1.000\,67 & 0.005\,90 \\ -0.050\,12 & 0.005\,90 & 1.051\,95 \end{pmatrix}.$$

However, if the average is taken for the four h.p.v. around the $(\bar{1}01)_p$ pole, the result is given as follows, after diagonalization.

$$P_1(\text{average}) = 1/4[P_1\{1'(+)\} + P_1\{2(+)\} + P_1\{5(+)\} + P_1\{6'(+)\}]$$
$$= \begin{pmatrix} 1.001\,90 & 0 & 0 \\ 0 & 1.000\,67 & 0 \\ 0 & 0 & 0.996\,26 \end{pmatrix},$$

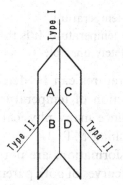

Fig. 1.15. The basic morphology (parallelogram morphology) of self-accommodation of martensites. (After Murakami *et al.*[33])

which is very close to the identity matrix. Thus, we can conclude that the above self-accommodation is very effective in reducing strains upon MT. Furthermore, we can show that even the self-accommodation consisting of two h.p.v. is effective, if they are type I twin-related or type II twin-related, although the compound twin-related pair is not (see Ref. [33]). In fact the spear-like morphology in Fig. 1.3 corresponds to the type I twin-related case.[21] From the analysis, we propose a parallelogram morphology as a basic morphology, as shown in Fig. 1.15,[33] which consists of type I and type II twinning only. The observed morphology in Fig. 1.13 is consistent with this morphology. However, a few short compound twin boundaries are also present in the figure. We believe that they were introduced accidentally as a result of impingement of h.p.v., which were independently nucleated, or from the unequal size of each h.p.v. In the above, we showed the self-accommodation for a specific case of 7R(14M) martensite, but the above mechanism is more common to other types of MTs, as Saburi and Wayman proposed, who carried out the above types of analysis extensively for various alloys.[29,34,35] The only difference between the two lies in the basic morphology. They proposed a diamond type morphology, which consists of type I twinning and compound type twinning.[29] However, the above parallelogram morphology is more consistent with the observations, and it is more reasonable from a theoretical point of view, as described above.

1.3 Martensitic transformations: thermodynamic aspects

Upon describing the thermodynamic aspects of martensitic transformations, we first define the transformation temperatures as follows.

M_s martensite start temperature
M_f martensite finish temperature

A_s reverse transformation start temperature
A_f reverse transformation finish temperature. It is the temperature above which the
    martensite becomes completely unstable.

These transformation temperatures can be determined by measuring some
physical properties as a function of temperature, as typically shown for a
Fe–Ni alloy in Fig. 1.16,[37] since many physical properties often drastically
change upon starting or finishing MT.

Since the martensitic transformations are not associated with a composi-
tional change, the free energy curves of both parent and martensite phases as a
function of temperature may be represented as schematically shown in Fig.
1.17, where T_0 represents the thermodynamic equilibrium temperature be-
tween the two phases, and $\Delta G^{p\rightarrow m}|_{M_s} = G^m - G^p$ represents the driving force
for the nucleation of martensite, where G^m and G^p represent the Gibbs free
energy of martensite and parent respectively. (It is customary to take the sign
of the driving force in an opposite sense,[37] but we use the above definition, so
as to be consistent with the sense of the free energy change upon MT.) The
same argument applies for the reverse transformation. Thus, T_0 was approxi-
mated by $1/2(M_s + A_s)$.[37] Now, a Gibbs free energy change of a system upon
MT may be written as follows:[37]

$$\Delta G = \Delta G_c + \Delta G_s + \Delta G_e = \Delta G_c + \Delta G_{nc}, \qquad (1.19)$$

where ΔG_c is a chemical energy term originating in the structural change from
parent to martensite, ΔG_s is a surface energy term between parent and martensite, ΔG_e is an elastic (plus plastic in the non-thermoelastic case, to be described
later) energy term around the martensite, and $\Delta G_{nc} = \Delta G_s + \Delta G_e$ is a non-

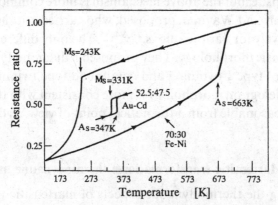

Fig. 1.16. Electrical resistance changes during cooling and heating Fe–Ni and Au–Cd
alloys, illustrating the hysteresis of the martensitic transformation on cooling, and the
reverse transformation on heating, for non-thermoelastic and thermoelastic transformations respectively. (After Kaufman and Cohen[37])

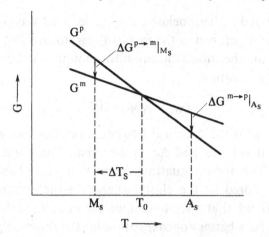

Fig. 1.17. Schematic representation of free energy curves for both parent and martensitic phases, and their relation to the M_s and A_s temperatures. ΔT_s is the supercooling required for the transformation.

chemical energy term. In most MTs, ΔG_{nc} is equally as large as ΔG_c, which is an essential point when we discuss MTs. Because of this, supercooling of ΔT_s is necessary for the nucleation of a martensite, and superheating is necessary for the reverse transformation in Fig. 1.17. By the same token, M_s is not the same as M_f, since the elastic energy around the martensite resists the growth of the martensite unless a further driving force (i.e. cooling) is given.

When we discuss MTs, we may classify them into two categories, thermoelastic and non-thermoelastic. These are typified in Fig. 1.16. In the case of the Au–47.5 at% Cd alloy, the transformation hysteresis is as small as 15 K, while in Fe–30 mass% Ni alloy, it is as large as ~ 400 K. In the former, the driving force for the transformation is very small (as evidenced by the small temperature hysteresis), the interface between parent and martensite is very mobile upon cooling and heating, and the transformation is crystallographically reversible in the sense that the martensite reverts to the parent phase in the original orientation. This type of transformation is called thermoelastic. On the other hand, in the latter, the driving force is very large, the interface between parent and martensite is immobile once the martensite grows to some critical size, and the reverse transformation occurs by the renucleation of the parent phase,[38] and thus the reverse transformation is not reversible. It is known that SME and SE are generally characteristic of thermoelastic transformations.[39]

Originally the notion of the thermoelastic transformation was introduced by Kurdjumov and Khandros,[40] who observed the growth and shrinkage of a martensite plate upon cooling and heating respectively, in a Cu–Al–Ni alloy.

Thus, they considered a thermoelastic equilibrium between chemical energy and elastic energy, which resists the forward transformation. Olson and Cohen[41] considered the thermoelastic equilibrium more quantitatively, and derived the following equation.

$$\Delta g_c + 2\Delta g_e = 0, \tag{1.20}$$

where $\Delta g_c = g^m - g^p$ is the chemical free energy change between parent and martensite per unit volume, and Δg_e is the elastic strain energy around the martensite plate. The above equation means that half of the chemical free energy change is stored as the elastic energy in the specimen. From this analysis, they showed that the A_s temperature could be below T_0. Thus, $T_0 = 1/2(M_s + A_f)$ is a better approximation for the thermoelastic transformation, as proposed by Tong and Wayman.[42]

We now discuss the effect of stress on MT, following Patel and Cohen's analysis.[43] As described in the previous section, MT proceeds by a shear-like mechanism, and thus interacts with an applied stress. Whether the stress assists or opposes the transformation is easily determined by calculating the work done on the system by the applied stress. Obviously, if the work is positive, the stress assists the transformation, and vice versa.

The work done on a system by an applied stress which produces the shape strain is,

$$\Delta G^s = m_1^p \tau + m_1^n \sigma_n. \tag{1.21}$$

Here τ is the shear stress along the habit plane in the direction d_1^p, and σ_n is the normal stress perpendicular to the habit plane. (For the notation of m_1^p and m_1^n, see Fig. 1.11.) By convention σ_n is positive for tensile stress and negative for compressive stress. $m_1^n = \Delta V/V$ is negative in most MTs, except for typical ferrous martensites. Thus, the sign of the second term on the right-hand side of Eq. (1.21) depends upon the sign of the stress, while the sign of the first term is always positive. This is because an h.p.v. is chosen among many possible variants (24 in general) such that the shape strain accommodates the applied stress most effectively. Thus, a shear stress always assists the transformation, but a normal stress may assist or oppose it, depending upon the sign of the stress and the volume change associated with the transformation. For hydrostatic pressure p $(p > 0)$ where the shear component is absent, Eq. (1.21) becomes $\Delta G^s = -p m_1^n$. Thus, the sign of ΔG^s depends upon that of m_1^n, in contrast to the uniaxial case. In fact, it is experimentally known that hydrostatic pressure aids the transformation for Au–Cd $(m_1^n < 0)$,[44] and opposes it for a Fe–Ni alloy $(m_1^n > 0)$,[43] consistent with the above equation.

An alternative way to analyze the effect of stress on MT is the use of the

Clausius–Clapeyron relationship. The relation for a uniaxial stress is written as follows,[45]

$$\frac{d\sigma}{dT} = -\frac{\Delta S}{\varepsilon} = -\frac{\Delta H^*}{\varepsilon T},\qquad(1.22)$$

where σ is a uniaxial stress, ε a transformation strain, ΔS the entropy of transformation per unit volume, and ΔH^* the enthalpy of the transformation per unit volume. We will use this equation in later sections. Strictly speaking, this equation applies for the equilibrium temperatures, but may be applied for the M_s temperatures as well, if the driving force to start the transformation is independent of temperature and stress.

References

1. L. C. Chang and T. A. Read, *Trans. AIME*, **189** (1951) 47.
2. W. J. Buehler, J. W. Gilfrich and R. C. Wiley, *J. Appl. Phys.*, **34** (1963) 1475.
3. M. W. Burkart and T. A. Read, *Trans. AIME*, **197** (1953) 1516.
4. Z. S. Basinski and J. W. Christian, *Acta Metall.*, **2** (1954) 101.
5. E. Hornbogen and G. Wassermann, *Z. Metallkd.*, **47** (1956) 427.
6. C. W. Chen, *J. Metals*, (Oct., 1957) 1202.
7. C. M. Wayman and J. D. Harrison, *J. Metals*, (Sept., 1989) 26.
8. *International Tables for X-ray Crystallography*, (The Kynotch Press, Birmingham, 1969) 15.
9. C. M. Wayman, *Introduction to Crystallography of Martensitic Transformations*, (MacMillan, NY, 1964) 23.
10. E. C. Bain, *Trans. AIME*, **70** (1924) 25.
11. Z. Nishiyama and S. Kajiwara, *Jpn. J. Appl. Phys.*, **2** (1963) 278.
12. J. Kakinoki, *Chemistry Suppl.* (*Kagaku Dojin*), **35**, Chapt. 5, (1968) 190 [in Japanese].
13. J. Kakinoki, S. Mifune, E. Kodera and T. Aikami, *J. Phys. Soc. Jpn.*, **32** (1972) 288.
14. H. Sato, T. S. Toth and G. Honjo, *J. Phys. Chem. Solids*, **28** (1967) 137.
15. K. Otsuka, T. Ohba, M. Tokonami and C. M. Wayman, *Scr. Metall. et Mater.*, **29** (1993) 1359.
16. Z. Nishiyama, *Martensitic Transformation*, (Academic Press, NY, 1978).
17. H. Warlimont and L. Delaey, *Prog. Mater. Sci.*, (Pergamon Press, Oxford, 1974).
18. R. W. Cahn, *Acta Metall.*, **1** (1953) 49.
19. N. D. H. Ross and A. G. Crocker, *Scr. Metall.*, **3** (1969) 37.
20. B. A. Bilby and A. G. Crocker, *Proc. Roy. Soc. Ser.* **A288** (1965) 240.
21. K. Otsuka, *Proc. Int. Conf. on Martensitic Transformations* (*ICOMAT*-86), (Nara, 1986) 35.
22. J. W. Christian, *The Theory of Transformations in Metals and Alloys*, (Pergamon Press, Oxford, 1965).
23. J. W. Christian and S. Mahajan, *Prog. Mater. Sci.*, **39**, No. 1/2, (Pergamon Press, Oxford, 1995).
24. M. S. Wechsler, D. S. Lieberman and T. A. Read, *Trans. AIME*, **197** (1953) 1503.
25. D. S. Lieberman, M. S. Wechsler and T. A. Read, *J. Appl. Phys.*, **26** (1955) 473.
26. J. S. Bowles and J. K. Mackenzie, *Acta Metall.*, **2** (1954) 129, 138, 224.
27. J. W. Christian, *J. Inst. Metals*, **84** (1955–6) 386.
28. B. A. Bilby and J. W. Christian, *Inst. Met. Monograph*, No. 18, (1955) 121.
29. T. Saburi and S. Nenno, *Proc. Int. Conf. on Solid–Solid Phase Transformations*, (AIME, NY, 1982) 1455.

30. K. Okamoto, S. Ichinose, K. Morii, K. Otsuka and K. Shimizu, *Acta Metall.*, **34** (1986) 2065.
31. K. Otsuka, *Materials Science Forum* (Proc. ICOMAT-89), **56–58** (1990) 393.
32. K. Otsuka and K. Shimizu, *Int. Metals Rev.*, **31**, No. 3, (1986) 93.
33. Y. Murakami, K. Otsuka, S. Hanada and S. Watanabe, *Mater. Sci. & Eng.*, **A189** (1994) 191.
34. T. Saburi and C. M. Wayman, *Acta Metall.*, **27** (1979) 979.
35. T. Saburi, C. M. Wayman, K. Takata and S. Nenno, *Acta Metall.*, **28** (1980) 15.
36. K. Adachi, J. Perkins and C. M. Wayman, *Acta Metall.*, **34** (1986) 2471.
37. L. Kaufman and M. Cohen, *Prog. Met. Phys.*, **7** (Pergamon Press, Oxford, 1957) 169.
38. H. Kessler and W. Pitsch, *Acta Metall.*, **15** (1967) 401.
39. K. Otsuka and K. Shimizu, *Scr. Metall.*, **4** (1970) 467.
40. G. V. Kurdjumov and L. G. Khandros, *Dokl. Nauk SSSR*, **66** (1949) 211.
41. G. B. Olson and M. Cohen, *Scr. Metall.*, **9** (1975) 1247.
42. H. C. Tong and C. M. Wayman, *Acta Metall.*, **22** (1974) 887.
43. J. R. Patel and M. Cohen, *Acta Metall.*, **1** (1953) 531.
44. Y. Gefen, A. Halway and M. Rosen, *Phil. Mag.*, **28** (1973) 1.
45. P. Wollants, M. De Bonte and J. R. Roos, *Z. Metallkd.*, **70** (1979) 113.
46. K. Otsuka and C. M. Wayman, in *Deformation Behavior of Materials*, ed. P. Feltham, Vol. II (Freund Publishing House, Israel, 1977) 91.
47. S. Ichinose, Y. Funatsu and K. Otsuka, *Acta Metall.*, **33** (1985) 1613.
48. K. A. Bywater and J. W. Christian, *Phil. Mag.*, **25** (1972) 1249.
49. K. M. Knowles and D. A. Smith, *Acta Metall.*, **29** (1981) 101.
50. O. Matsumoto, S. Miyasaki, K. Otsuka and H. Tamura, *Acta Metall.*, **35** (1987) 2137.

2

Mechanism of shape memory effect and superelasticity

K. OTSUKA AND C. M. WAYMAN

Figure 2.1[1] shows stress–strain (S–S) curves as a function of temperature for a Cu–34.7 mass% Zn–3.0 mass% Sn alloy single crystal. When it is tensile tested at a temperature below A_f, strains remain, but they recover by heating to a temperature above A_f, as indicated by dotted lines. This is the shape memory effect. On the contrary, when it is tensile tested at a temperature above A_f, strains recover just by unloading. This is superelasticity. In this chapter, we will present experimental results mostly obtained with single crystals, since they are more suited for detailed analysis. The results for polycrystals will be presented in later chapters in detail. We will discuss superelasticity first, and then the shape memory effect, since this is the order which is easier to follow for the readers.

2.1 Stress-induced martensitic transformation and superelasticity

2.1.1 Stress-induced transformations and associated superelasticity

In Section 1.3, we learned that a uniaxial stress always assists MT. Thus, we can expect the stress-induced martensitic transformation (SIM) above M_s. This is shown in a series of S–S curves for a Cu–14.1 mass% Al–4.0 mass% Ni alloy single crystal in Fig. 2.2.[2] In this experiment, tensile tests were carried out from low temperatures to higher ones in order, and the applied stresses were restricted as small as possible to eliminate the effect of the strains in the previous tests. We observe two types of S–S curves in the figure; one with a sharp peak, and the other with a smooth curve with a very small stress hysteresis. The former corresponds to the β_1 (DO$_3$ type ordered structure)–γ'_1 (2H) transformation, and the other to the β_1–β'_1 (18R) transformation. If we plot the critical stresses in Fig. 2.2(a) as a function of temperature, we obtain

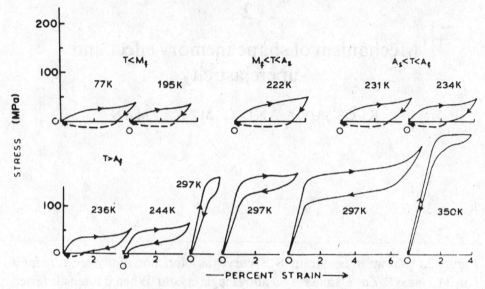

Fig. 2.1. Tensile stress–strain curves as functions of temperature for
Cu–34.7 mass% Zn–3.0 mass% Sn alloy single crystal, for which $M_s = 221$ K,
$M_f = 208$ K, $A_s = 223$ K and $A_f = 235$ K. (After Eisenwasser and Brown[1])

two straight lines with different slopes as shown in Fig. 2.2(b). The different
slopes correspond to the different transformations described above. The reason why the critical stress increases with increasing temperature is that the
parent phase is more stable in higher temperature ranges, and thus higher
stress is required for SIM. This is in accord with the Clausius–Clapeyron
relationship in Eq. (1.22) in Section 1.3, since ΔH^* is exothermic for SIM in the
equation. In the following, we will focus attention on the latter transformation
β_1–β_1', since it behaves better, without the scattering of data compared with the
former, for which a rather large scattering is observed in Fig. 2.2(b). In Fig.
2.2(a), we observe nonlinear superelastic loops. We show the superelastic loop
in the full strain, along with the corresponding micrographs in Fig. 2.3,[3] which
were taken from a different specimen. From this figure, we can immediately
notice that superelasticity is realized by the SIM upon loading, and RT upon
unloading. Superelasticity is usually realized when a specimen is stressed
above A_f, since the stress-induced martensite is stable only under stress, and it
is unstable in the absence of stress at temperatures above A_f.

In Fig. 2.3 only one h.p.v. is stress-induced among 24 h.p.v. It is generally
accepted that an applied stress interacts with the shape strain, and the h.p.v. is
selected, for which the Schmid factor (sin χ cos λ) for the shear component of
the shape strain (d_1^p) is the maximum.[4] Here χ is an angle between the habit
plane and the tensile axis, and λ an angle between the direction of the shear

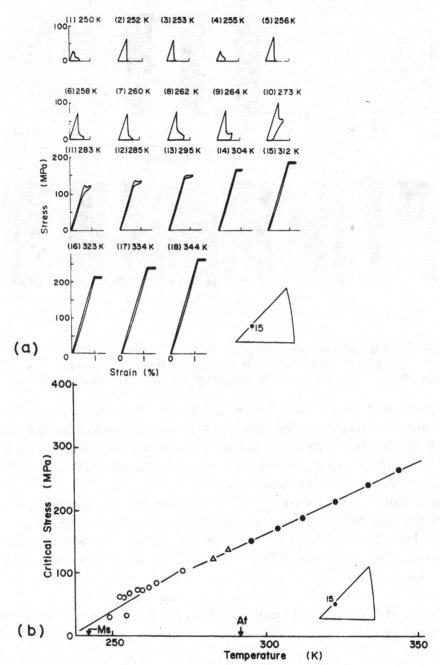

Fig. 2.2. (a) Stress–strain curves as functions of temperature for
Cu–14.1 mass% Al–4.0 mass% Ni alloy single crystal, for which $M_s = 242\,\text{K}$,
$M_f = 241\,\text{K}$, $A_s = 266\,\text{K}$ and $A_f = 291\,\text{K}$. (b) Critical stress *vs* temperature plots for
the same specimen. The symbol (○) corresponds to β_1–γ_1' transformation, the symbol
(•) to β_1–β_1', and the symbol (△) to that involving both transformations. (After
Horikawa *et al.*[2])

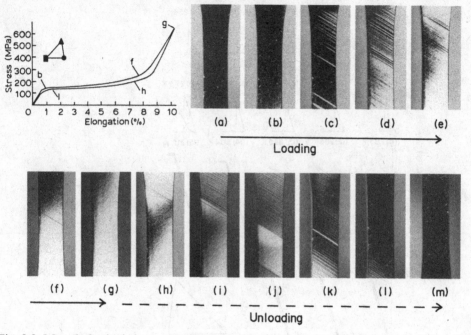

Fig. 2.3. Morphological changes associated with $\beta_1-\beta_1'$ stress-induced transformation upon loading, and its reverse transformation upon unloading for Cu–14.2 mass% Al–4.2 mass% Ni alloy single crystal. (After Otsuka *et al.*[3])

component of the shape strain and the tensile axis. This is especially true for the above $\beta_1-\beta_1'$ transformation, although there are some objections in some ferrous alloys.[5] As will be shown below, the above variant selection rule is approximately equivalent to saying that the variant is selected, which gives the largest transformation strain under stress.[7]

It is possible to assess the entropy of transformation (ΔS) or the enthalpy of transformation (ΔH^*) by using the Clausius–Clapeyron relationship, since the left-hand side ($d\sigma/dT$) of Eq. (1.22) is obtained from a critical stress vs. temperature curve as in Fig. 2.2(b), and the transformation strain may be obtained by experiment such as in Fig. 2.4 or by a calculation, as given in the next section. Those values for the $\beta_1-\beta_1'$ transformation in Fig. 2.2 were $\Delta S = -1.35$ J K^{-1}/mol and $\Delta H^* = -310$ J/mol.[2] See Refs. [4,6] and Table 2.1 on SE in various alloys.

2.1.2 Orientation dependence and calculation of superelastic strain

Figure 2.4[2] shows the orientation dependence of the superelastic strain (or transformation strain), which corresponds to the plateau length in the S–S curves as a function of orientation. This orientation dependence can be ex-

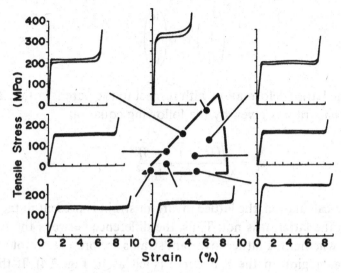

Fig. 2.4. Orientation dependence of stress–strain curves for β_1–β_1' transformation for Cu–Al–Ni single crystal. (After Horikawa *et al.*[2])

plained by actually calculating the transformation strain for uniaxial stress. There are two ways of calculating it, either by using the shape strain or by using the lattice deformation.

(1) The calculation based on the shape strain[3]

In this case, the transformation strain is given by the following equation.

$$\varepsilon_c = \sqrt{(m_1^p \sin \chi)^2 + 2m_1^p \sin \chi \cos \lambda + 1} - 1 + m_1^n \sin^2 \chi. \tag{2.1}$$

This equation is similar to the Schmid–Boas relation describing the relation between shear strain and elongation for slip, but the normal component of the shape strain is also taken into account in the above equation. (For the meaning of the symbols, see Fig. 1.11.) In thermoelastic alloys, usually $m_1^n (\Delta V/V) \ll m_1^p$ and $(m_1^p \sin \chi)^2 \ll 1$. Thus,

$$\varepsilon_c \approx m_1^p \sin \chi \cos \lambda. \tag{2.2}$$

This means that ε_c is proportional to the Schmid factor for the shape strain. This is what we have stated in the previous section with respect to the variant selection.

(2) The calculation based on the lattice deformation $\boldsymbol{B}$[7]

According to the lattice deformation, any vector $[uvw]$ is changed into $[UVW]$ upon MT such that

$$\begin{pmatrix} U \\ V \\ W \end{pmatrix} = \boldsymbol{B} \begin{pmatrix} u \\ v \\ w \end{pmatrix}, \tag{2.3}$$

where $\boldsymbol{B}$ is the lattice deformation with respect to the parent lattice. Thus, the transformation strain is given by the following equation.

$$\varepsilon_c = \sqrt{\frac{U^2 + V^2 + W^2}{u^2 + v^2 + w^2}} - 1. \tag{2.4}$$

In the former calculation, the lattice invariant strain is present in the martensite, while in the latter, it is not. Thus, the difference between the two treatments lies in whether the lattice invariant strains are present or not at the end of the plateau region in the S–S curve (such as in Fig. 2.4). If the lattice invariant strain does not move under the stress, the first calculation should hold, while if the stress-induced martensite becomes a single crystal by detwinning for example, the second calculation should hold. Thus, the former calculation gives the minimum transformation strain, while the latter gives the maximum. Thus, which calculation is better depends upon the alloy system and stress level etc. In the above case, the former calculation fitted better, as shown in Fig. 2.5.[2]

As will be shown in Table 2.1 in Section 2.2.3, many SMAs are β-phase alloys, which have strong elastic anisotropy. Thus, the Young's modulus in the parent phase was calculated for the above alloy as a function of orientation in Fig. 2.6.[2] We observe that the Young's modulus is 10 times lower in $\langle 001 \rangle$ orientation than that in $\langle 111 \rangle$. Because of this, it is possible to obtain an

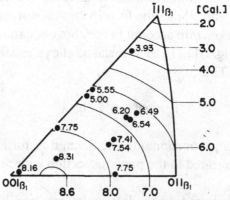

Fig. 2.5. Comparison of observed transformation strains (•) (in %) to those calculated from shape strain (contour lines) for the β_1–$\beta_1{}'$ transformation in Fig. 2.4. (After Horikawa *et al.*[2])

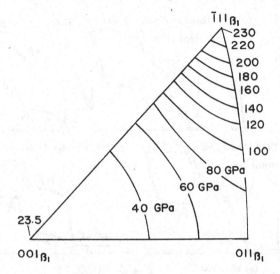

Fig. 2.6. Orientation dependence of Young's modulus calculated for the β_1 parent phase in Cu–Al–Ni alloy. (After Horikawa *et al.*[2])

unusually large *elastic* strain in this alloy, as will be discussed in the next section.

The effect of the strain rate on superelastic loops may be summarized as follows.[3] The stress hysteresis is usually found to increase with increasing strain rate for the following reasons. If we take a midpoint between loading and unloading curves as an equilibrium stress between parent and martensite, half of the stress hysteresis becomes the driving force for the forward and reverse MT. Thus, for a higher strain rate, we need a larger driving force, which results in a larger stress hysteresis. For more details, see Ref. [3].

2.1.3 Successive stress-induced transformations

In Section 1.3, SIM was explained by considering the equivalence of changing stress and temperature from an energetic point of view. However, stress and temperature are independent thermodynamic variables, in principle. Thus, two effects may be different in certain cases. We saw such an example in Fig. 2.2, where two distinct transformations appeared depending upon test temperature and stress. Furthermore, it is also possible that various martensites appear in temperature–stress space, including a MT from one martensite to another (a martensite-to-martensite transformation). This is typically shown in Fig. 2.7,[8] which represents a series of S–S curves for a Cu–14.0 mass% Al–14.2 mass% Ni alloy single crystal in $\langle 001 \rangle$ orientation. In the S–S curves, each

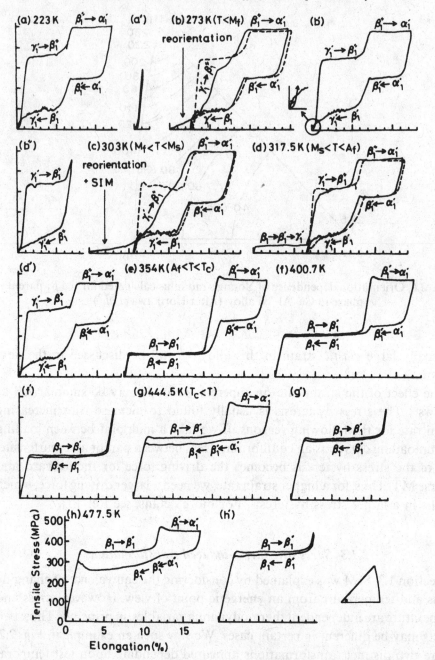

Fig. 2.7. Stress–strain curves as functions of temperature representing successive stress-induced transformations in a Cu–14.0 mass% Al–4.2 mass% Ni alloy single crystal. (After Otsuka et al.[8])

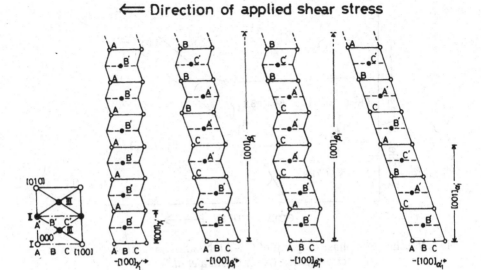

Fig. 2.8. Crystal structures of various stress-induced martensites in a Cu–Al–Ni alloy single crystal; open circles represent Al atoms and closed ones Cu (and extra Ni and Al) atoms. (After Otsuka *et al.*[8])

stage represents a SIM or its RT, which was confirmed by neutron diffraction under stress. The structure of each martensitic phase determined by the neutron diffraction is shown in Fig. 2.8,[8] and the structure of the parent phase β_1 is a DO_3 type ordered structure. It is interesting to note that all the martensite structures are long period stacking order structures with the same basal plane, and the only difference in their structures lies in their stacking sequence alone. If we plot the critical stress for each transformation, we obtain the critical stress vs. temperature curves, from which we can draw a phase diagram as shown in Fig. 2.9.[8] In the figure, we notice two martensitic phases β_1' and β_1'', which have different structures, as shown in Fig. 2.8. This means that when a martensite is stress-induced from γ_1', β_1'' martensite is obtained, while when it is stress-induced from β_1, β_1' martensite is obtained. We think that β_1' is a stable phase and β_1'' is a metastable one which is realized in connection with the transformation mechanism. There are many interesting points on these martensite-to-martensite transformations including the above problem, but these have been omitted because of space limitation. See Refs. [6,8] for more details.

It is interesting to note that as much as 18% superelastic strain is available in Fig. 2.7 (e–g). It is also surprising that 2% *elastic strain* in the parent phase is available at a high temperature in (h). This becomes possible, since the Young's modulus is very low in ⟨001⟩ orientation (see Fig. 2.6) and the fracture stress is

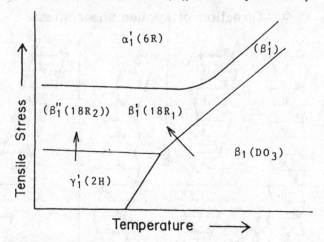

Fig. 2.9. Schematic phase diagram of Cu–Al–Ni alloy in temperature–stress coordinates. (After Otsuka *et al.*[8])

high in this alloy. This property may lead to some interesting engineering applications as well as superelasticity.

Although the martensite-to-martensite transformations were described only for a Cu–Al–Ni single crystal, similar transformations were also found in Cu–Zn,[9] Cu–Zn–Al,[10,11] Au–Ag –Cd,[12] Au–47.5 at% Cd[13] and Au–49.5 at% Cd[14,15] alloy single crystals.

2.2 Shape memory effect

2.2.1 Origin and mechanism of shape memory effect

As shown in Fig. 2.1, the shape memory effect is a phenomenon such that even though a specimen is deformed below A_s, it regains its original shape by virtue of the reverse transformation upon heating to a temperature above A_f. The deformation may be of any kind such as tension, compression or bending etc. (e.g. see Fig. 2.10 (a–c)),[16] as long as the strain is below some critical value, as will be discussed in the next section. The origin is of course the presence of the reverse transformation upon heating. As stated above, SME occurs when specimens are deformed below M_f or at temperatures between M_f and A_s, above which the martensite becomes unstable. Depending upon the temperature regimes, the mechanisms of SME are slightly different. In the following, we will describe the former case ($T \leq M_f$) first, by using a simplified model of a single crystal parent phase, as shown in Fig. 2.11.[17] Suppose we cool a single crystal parent phase (a) to a temperature below M_f. Then, martensites are

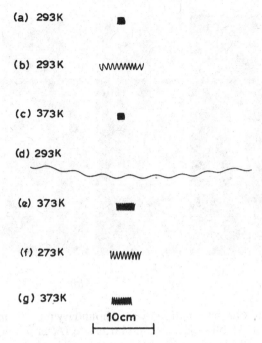

(a) 293K

(b) 293K

(c) 373K

(d) 293K

(e) 373K

(f) 273K

(g) 373K

10cm

Fig. 2.10. Demonstration of shape memory effect (a–c) and two-way shape memory effect (d–g) in Ti–50.0 at% Ni alloy. (After Otsuka and Shimizu[16])

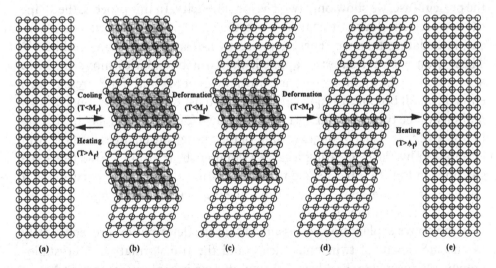

Fig. 2.11. Mechanism of shape memory effect; (a) original parent single crystal, (b) self-accommodated martensite, (c–d) deformation in martensite proceeds by the growth of one variant at the expense of the other (i.e. twinning or detwinning), (e) upon heating to a temperature above A_f, each variant reverts to the parent phase in the original orientation by the reverse transformation. (After Otsuka[17])

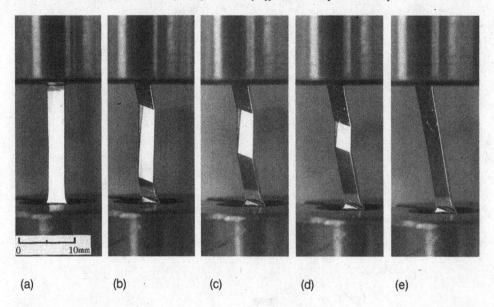

(a)   (b)   (c)   (d)   (e)

Fig. 2.12. Demonstrates the deformation by twin boundary movement [$\langle 111 \rangle_m$ type II twinning] in Cu–Al–Ni single variant martensite. (After Ichinose *et al.*[19])

formed in a self-accommodating manner (b) as described in Section 1.2.5. In the present case, we show only two c.v. for simplicity. In this process, the shape of the specimen does not change, since the transformation occurs in a self-accommodating manner. These c.v. are twin-related and quite mobile.[7,18] Thus, if an external stress is applied, the twin boundaries move so as to accommodate the applied stress, as shown in (c) or (d), and if the stress is high enough, it will become a single variant of martensite under stress. Such a high mobility of the twin boundary is typically shown for a γ_1' Cu–Al–Ni martensite in Fig. 2.12,[19] in which a single variant of the martensite changes into the twin orientation by shear, creating a large twinning shear strain. Now, when the specimen in Fig. 2.11(d) is heated to a temperature above A_f, RT occurs, and if the RT is crystallographically reversible, the original shape is regained as in (e). This is the mechanism of SME.

In the above explanation, we assumed that the deformation proceeds solely by the movement of twin boundaries, and the transformation is crystallographically reversible. If either of the conditions is broken, complete SME is not obtained. This point will be discussed in the next section. It is also obvious that the memorized shape is that of the parent phase only. This point will be discussed in Section 2.2.4 also. In the above explanation, we simplified the situation greatly. In reality, we have learned in Section 1.2.5 that the formed martensites are self-accommodated by two or four h.p.v., and each h.p.v.

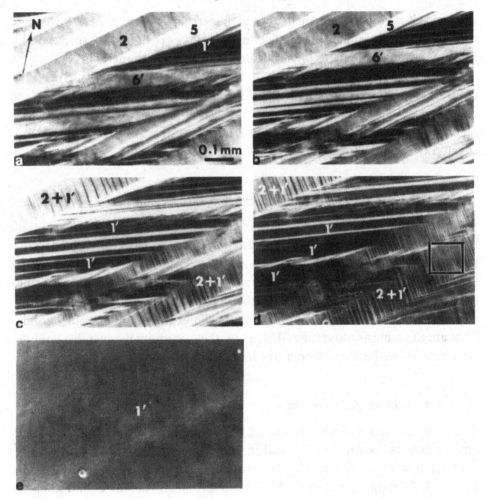

Fig. 2.13. Optical microscope observation of variant coalescence in a Cu–Zn–Ga single crystal; (a) self-accommodated martensite, (b–e) deformation by twin boundary movement into a single variant martensite 1′, which is the favorable variant under the stress. (After Saburi *et al.*[20])

contains twins (or dislocations) as a lattice invariant strain. Thus, apart from the case of dislocations, the martensites of a specimen concerned are twin-related to each other. Thus, the same principle applies for such a complicated real case. That is, the particular c.v. is favored under the applied stress (i.e. the c.v. which gives the largest transformation strain under the stress. See Section 2.1.2). That is, the particular variant grows at the expense of the others until the specimen becomes the particular single variant, if the stress is high enough. Such a deformation process of the self-accommodated martensites is clearly shown for a Cu–Zn–Ga single crystal in Fig. 2.13.[20] In the figure, the martensites are self-accommodated in (a), but with increasing stress, the c.v. 1' which is a favorable variant under the applied tensile stress, grows at the expense of the others, and finally the specimen becomes the single variant 1' in (e).

When a specimen is deformed at temperatures $M_f < T < A_s$, SIM also contributes to the deformation, in addition to the above process of variant coalescence. However, the selected variant is the one which is favored by the applied stress, as discussed in Section 2.1.2. Thus, the mechanism of SME is essentially the same as above.

2.2.2 *Calculation of maximum recoverable strain*

The calculation of the recoverable strain by SME of a single crystal is essentially the same as that for SE, for the reason described below. As described in the previous section, the shape of a specimen does not change upon MT due to self-accommodation, i.e. it is the same as that in the parent phase. Meanwhile, the end result of the deformation in the martensitic state is a single variant, which is most favored by the applied stress. Thus, the initial and final conditions are the same as those for a SIM. Thus, the maximum recoverable strain is obtained by the lattice deformation matrix[7] in the same way as for SIM.

2.2.3 *Conditions for good shape memory and superelastic characteristics*

So far we have discussed SE and SME separately, but they are closely related phenomena, as seen in Fig. 2.1 and in later chapters. Thus, we discuss both together in this section. The relation between the two is schematically shown in Fig. 2.14.[16] In principle, both SME and SE are observable in the same specimen, depending upon the test temperature, as long as the critical stress for slip is high enough, as will be discussed below. SME occurs below A_s, followed by heating above A_f, while SE occurs above A_f, where martensites are completely unstable in the absence of stress. In the temperature regime between A_s and A_f, both occur partially. In Fig. 2.14, a straight line with a positive slope represents the critical stress to induce a martensite, following the Clausius–Clapeyron relationship. The straight lines with negative slopes (A or B) represent the critical stress for slip. Since slip never recovers upon heating or unloading, the stress must be below the line to realize SME or SE. It is quite clear that no SE is realized, if the critical stress is as low as the line B, since slip occurs prior to the onset of SIM.

It is now clear from the above discussions that the essential conditions for the realization of SME and SE are the crystallographic reversibility of MT and the avoidance of slip during deformation.[18,21,22] It was once proposed that

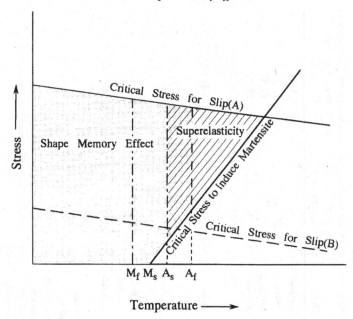

Fig. 2.14. Schematic diagram representing the region of shape memory effect and superelasticity in temperature–stress coordinates; (A) represents the critical stress for the case of a high critical stress and (B) represents the critical stress for a low critical stress.

SME is characteristic of thermoelastic MT and of ordered alloys.[21] Table 2.1 shows the list of non-ferrous SMAs, which indicates that they are all thermoelastic and ordered, except for In–Tl and Mn–Cu etc. exhibiting an FCC–FCT transformation associated with extremely small transformation strains. Ferrous alloys, which are listed in Table 5.1, are generally non-thermoelastic (except for an ordered Fe–Pt alloy[23] etc.) and are different in SME characteristics, and they will be discussed separately. The reason why thermoelastic alloys are favored for SME and SE is due partly to the small driving force for the transformation (which is evidenced by a small temperature hysteresis), which avoids the introduction of dislocations, and partly to the presence of many mobile twins, thus leading to crystallographic reversibility. The ordered structure is also related to crystallographic reversibility and the avoidance of slip. The former is explained in Fig. 2.15 for the orthorhombic (ordered HCP)–B2 transformation.[22] If the alloy is disordered, there are three possible paths to the parent phase, but if it is ordered, there is only one path without destroying the original ordered parent structure upon RT. Since the wrong path changes the structure of the parent phase and thus increases the energy of the system, crystallographic reversibility is guaranteed, if it is ordered. The ordered structure is also favorable for the avoidance of slip, since

Table 2.1. *Non-ferrous alloys exhibiting perfect shape memory effect and superelasticity*

Alloy	Composition (at%)	Structure change	Temperature hysteresis (K)	Ordering	Ref. on structure	Ref. on crystallography, SME and SE
Ag–Cd	44–49 Cd	B2–2H	~ 15	ordered	26	27, 28
Au–Cd	46.5–48.0 Cd	B2–2H	~ 15	ordered	29, 30	31–34
	49–50 Cd	B2–trigonal	~ 2	ordered	35, 36	33, 37
Cu–Zn	38.5–41.5 Zn	B2–M (modified) 9R	~ 10	ordered	38	10, 39
Cu–Zn–X (X = Si,Sn,Al,Ga)	A few at%	B2–M9R	~ 10	ordered	38	9, 10, 11, 28, 40
Cu–Al–Ni	28–29 Al, 3.0–4.5 Ni	DO$_3$–2H	~ 35	ordered	41, 42	3, 8, 43–45
Cu–Sn	~ 15 Sn	DO$_3$–2H, 18R	—	ordered	46, 47	48
Cu–Au–Zn	23–28 Au, 45–47 Zn	Heusler–18R	~ 6	ordered	49	50
Ni–Al	36–38 Al	B2–3R, 7R	~ 10	ordered	51, 52	53–56
Ti–Ni	49–51 Ni	B2–monoclinic	~ 30	ordered	57	59–61
		B2–R-phase-(monoclinic)	~ 2	ordered	58	62, 63, 64
Ti–Ni–Cu	8–20 Cu	B2–orthorhombic-(monoclinic)	4–12	ordered	65	66
Ti–Pd–Ni*	0–40 Ni	B2–orthorhombic	30–50	ordered	67, 68	69–71
In–Tl	18–23 Tl	FCC–FCT	~ 4	disordered	72	73–75
In–Cd	4–5 Cd	FCC–FCT	~ 3	disordered	76	76
Mn–Cd	5–35 Cu	FCC–FCT	—	disordered	77	78

* Ti–Pd–Ni alloys with high Pd content do not exhibit good SME unless specially thermomechanically treated.

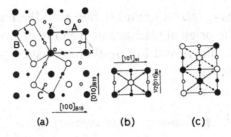

Fig. 2.15. (a) Three possible lattice correspondences in the reverse transformation from B19 type (ordered HCP) to B2 parent phase. (b) Atomic configuration obtained by the reverse transformation for variant A in (a). (c) Atomic configuration obtained by the reverse transformation for variant B in (a). Projection plane is the $(101)_{B2}$ plane; larger and smaller circles represent atom positions in alternate $(101)_{B2}$ planes; arrows indicated by solid and broken lines represent unlike and like atoms in the nearest neighbor positions, respectively. (After Otsuka and Shimizu[22])

the ordered structure usually has a higher critical stress for slip than that of a disordered one (because of superdislocations etc.). From these arguments, it is expected that SME and SE characteristics may be improved by hardening of thermoelastic alloys. From the viewpoint of physical metallurgy, there are three ways to increase the critical stress for slip, i.e. (1) solution hardening, (2) precipitation hardening, (3) work hardening. The improvement of SM characteristics has already been achieved in Ti–Ni alloys by using these principles.[24] These will be discussed in detail in Chapter 3. In–Tl and Mn–Cu are known to exhibit crystallographic reversibility, although they are not ordered. This is possibly because the transformation strains are so small. SME and SE as well as the structure of martensites in various alloys are found in the references, which are listed in Table 2.1. However, it lists only representative ones, and is not exhaustive.

There are some arguments that martensitic transformations in general are crystallographically reversible. It is true that if a MT is reversible, the strains are most efficiently released after RT. However, there is no such guarantee in non-thermoelastic alloys. It was shown in Fe–Ni alloys that RT occurred by the renucleation of the parent phase, and the original single crystal changed into polycrystals after RT.[25] It is also known that most ferrous martensites do not exhibit SME under normal heat-treatments, although some of them exhibit pretty good SME under certain thermomechanical treatments. Even in this case, however, they exhibit SME only when martensites are stress-induced, and they do not exhibit SME when they are deformed in the complete martensitic state (i.e. $T < M_f$) (see Chapter 5). Furthermore, in FCC–HCP transformations, the HCP martensites often transform back to the parent phase in twin orientation under stress rather than in the original orientation.

In ordered thermoelastic transformations, however, the martensites revert to the parent phase in the original orientation even under stress. In this respect, SME in thermoelastic alloys and in non-thermoelastic ones is different. See Chapter 5 for more details.

2.2.4 Two-way shape memory effect

In SME, only the shape of the parent phase is remembered, as was described in Section 2.2.1. However, it is possible to remember the shape of the martensitic phase under certain conditions. In Fig. 2.10, when the applied stress is small, the specimen reverts to the original shape completely by SME (a–c). However, when the applied stress is too large (d), irreversible slip occurs, and the shape does not revert to the original one even after heating above A_f [cf. (c) and (e)]. However, in the next cooling cycle, the specimen elongates automatically as shown in (f). Then, if heating and cooling is repeated, the specimen changes its shape between (g) and (f), respectively. The specimen now remembers the shape of (f) in the martensitic state. This is called the two-way shape memory effect (TWME). In contrast to this, the previous SME is sometimes called the one-way shape memory effect. The reason why the specimen remembered the shape of (f) may be explained as follows. Upon heavy deformation in (d), dislocations are introduced so as to stabilize the configuration of martensites. These dislocations exist even in the parent phase after RT upon heating, and the stress field around them induces particular h.p.v. upon cooling. There are several thermomechanical treatments, such as the introduction of plastic deformation,[79] constraint ageing,[80] thermal cycling,[81] utilization of precipitates[82,83] etc., but they will be discussed in detail in Chapters 3 and 7.

2.3 Rubber-like behavior

There is a very strange phenomenon called 'the rubber-like behavior'. This was found by Ölander[84] in a Au–47.5 at% Cd alloy (γ'_2 martensite) in the 1930s, but its origin is still not understood. When the alloy is deformed just after the MT, the deformation behavior is plastic and it exhibits SME. However, when the alloy is aged in the martensitic state for 14 hrs or so, it becomes pseudoelastic. As far as the S–S curve is concerned, it is very similar to superelasticity, but it is quite different from the latter in that it occurs solely in the martensitic state. The same behavior was later observed in thermoelastic alloys such as Au–Cu–Zn,[85] Cu–Al–Ni,[86] Cu–Zn–Al,[87] Au–49.5 at% Cd (ζ'_2 martensite)[88] and In–Tl[89,90] etc. It is known that the behavior occurs by the reversible movement of twin boundaries,[91,86] but the reason why the twin boundary becomes reversible after ageing is not understood. Since the origin and mechanism are not established, and space is limited, see Refs. [92–95] for more details. Just recently, a new mechanism was proposed to explain all aspects of the behavior. See the original paper.[115]

Terminology

There is some confusion over the terminologies of pseudoelasticity, superelasticity and rubber-like behavior. Thus, before closing this chapter, we define them as follows. When an apparent plastic deformation recovers just by unloading at a constant temperature (i.e. the S–S curve is characterized by a closed loop), it is called *pseudoelasticity*, irrespective of its origin. It is a generic term which encompasses both superelastic and rubber-like behavior. When a closed loop originates from a stress-induced transformation upon loading and the reverse transformation upon unloading, it is called *superelasticity*. If it occurs by the reversible movement of twin boundaries *in the martensitic state*, it is called *rubber-like behavior*. These definitions are consistent with the above descriptions in this chapter, and there should be no confusion.

To the authors' knowledge, there is only one book on shape memory alloys[96] *in English*, but there are many conference proceedings. These are listed at the end of the References [97–112], which are useful for further studies.

References

1. J. D. Eisenwasser and L. C. Brown, *Metall. Trans.*, **3** (1972) 1359.
2. H. Horikawa, S. Ichinose, K. Morii, S. Miyazaki and K. Otsuka, *Metall. Trans.*, **19A** (1988) 915.
3. K. Otsuka, C. M. Wayman, K. Nakai, H. Sakamoto and K. Shimizu, *Acta Metall.*, **24** (1976) 207.
4. K. Otsuka and C. M. Wayman, in *Deformation Behavior of Materials*, ed. P. Feltham, Vol. II, (Freund Publishing House, Israel, 1977) 98.
5. M. Kato and T. Mori, *Acta Metall.*, **24** (1976) 853.
6. K. Otsuka and K. Shimizu, *Proc. Int. Conf. on Solid–Solid Phase Transformations*, (AIME, NY, 1982) 1267.
7. T. Saburi and S. Nenno, *Proc. Int. Conf. on Solid–Solid Phase Transformations*, (AIME, NY, 1982) 1455.
8. K. Otsuka, H. Sakamoto and K. Shimizu, *Acta Metall.*, **27** (1979) 585.
9. W. Arneodo and M. Ahlers, *Scr. Metall.*, **7** (1973) 1287.
10. K. Takezawa, H. Sato, Y. Abe and S. Sato, *J. Jpn. Inst. Metals*, **43** (1979) 235.
11. H. Sato, K. Takezawa and S. Sato, *Sci. Rep. Res. Inst.*, Tohoku Univ., **29A**, Suppl. No. 1 (1985) 85. [in Japanese]
12. S. Miura, F. Hori and N. Nakanishi, *Phil. Mag.*, **40** (1979) 611.
13. H. Sakamoto and K. Shimizu, *J. de Phys.*, **43**, Suppl. No. 122 (ICOMAT-82) (1982) C4-623.
14. S. Miura, T. Mori, N. Nakanishi, Y. Murakami and S. Kachi, *Phil. Mag.*, **34** (1976) 337.
15. Y. Murakami, K. Morii, K. Otsuka and T. Ohba, *J. de Phys.* **IV** (Proc. ICOMAT-95) C8-1077.

16. K. Otsuka and K. Shimizu, *Int. Metals Rev.*, **31**, No. 3 (1986) 93.
17. K. Otsuka, Chapter 3, *Functional Metallic Materials*, ed. M. Doyama and R. Yamamoto, (University of Tokyo Press, 1985) 56. [in Japanese]
18. K. Otsuka, *Jpn. J. Appl. Phys.*, **10** (1971) 571.
19. S. Ichinose, Y. Funatsu and K. Otsuka, *Acta Metall.*, **33** (1985) 1613.
20. T. Saburi, C. M. Wayman, K. Takata and S. Nenno, *Acta Metall.*, **28** (1980) 15.
21. K. Otsuka and K. Shimizu, *Scr. Metall.*, **4** (1970) 467.
22. K. Otsuka and K. Shimizu, *Scr. Metall.*, **11** (1977) 757.
23. D. P. Dunne and C. M. Wayman, *Met. Trans.*, **4** (1973) 147.
24. S. Miyazaki, Y. Ohmi, K. Otsuka and Y. Suzuki, *J. de Phys.*, **43**, Suppl. 12 (Proc. ICOMAT-82), (1982) C4-255.
25. H. Kessler and W. Pitsch, *Acta Metall.*, **15** (1967) 401.
26. D. B. Masson and C. S. Barrett, *Trans. AIME*, **212** (1985) 260.
27. R. V. Krishnan and L. C. Brown, *Metall. Trans.*, **4** (1973) 423.
28. T. Saburi and C. M. Wayman, *Acta Metall.*, **28** (1980) 1.
29. A. Ölander, *Z. Kristallogr.*, **83A** (1932) 145.
30. T. Ohba, Y. Emura, S. Miyazaki and K. Otsuka, *Mater. Trans. JIM*, **31** (1990) 12.
31. D. S. Lieberman, M. S. Wechsler and T. A. Read, *J. Appl. Phys.*, **26** (1955) 473.
32. K. Morii, T. Ohba, K. Otsuka, H. Sakamoto and K. Shimizu, *Acta Metall.*, **33** (1992) 29.
33. N. Nakanishi, T. Mori, S. Miura, Y. Murakami and S. Kachi, *Phil. Mag.*, **28** (1973) 277.
34. S. Miura, T. Mori, N. Nakanishi, Y. Murakami and S. Kachi, *Phil. Mag.*, **34** (1976) 337.
35. T. Ohba, Y. Emura and K. Otsuka, *Mater. Trans. JIM*, **33** (1992) 29.
36. T. Tadaki, Y. Katano and K. Shimizu, *Acta Metall.*, **26** (1978) 883.
37. K. Morii, S. Miyazaki and K. Otsuka, *Proc. Int. Conf. on Martensitic Transformations (ICOMAT-92)*, (Monterey, 1992) 1125.
38. T. Tadaki, M. Tokoro and K. Shimizu, *Trans. JIM*, **16** (1975) 285.
39. T. A. Schroeder and C. M. Wayman, *Acta Metall.*, **28** (1977) 1375.
40. J. De Vos, L. Delaey and E. Aernoud, *Acta Metall.*, **26** (1979) 1745.
41. M. J. Duggin, *Acta Metall.*, **14** (1966) 123.
42. J. Ye, M. Tokonami and K. Otsuka, *Metall. Trans.*, **21A** (1990) 2669.
43. K. Oishi and L. C. Brown, *Metall. Trans.*, **2** (1971) 1971.
44. H. Tas, L. Delaey and A. Deruyttere, *Metall. Trans.*, **4** (1973) 2833.
45. V. V. Martynov and L. G. Khandros, *Dokl. Akad. Nauk USSR*, **233** (1977) 245.
46. Z. Nishiyama, K. Shimizu and H. Morikawa, *Trans. JIM*, **9** (1968) 307.
47. K. Shimizu, H. Sakamoto and K. Otsuka, *Trans. JIM*, **16** (1975) 581.
48. S. Miura, Y. Morita and N. Nakanishi, *Shape Memory Effect in Alloys*, ed. J. Perkins, (Plenum Press, NY, 1975) 389.
49. T. Tadaki, H. Okazaki, Y. Nakata and K. Shimizu, *Mater. Trans. JIM*, **31** (1990) 941.
50. S. Miura, S. Maeda and N. Nakanishi, *Phil. Mag.*, **30** (1974) 565.
51. S. Rosen and J. A. Goebel, *Trans. AIME*, **242** (1968) 722.
52. V. V. Martynov, K. Enami, L. G. Khandros, S. Nenno and A. V. Tkachenko, *Phys. Met. Metallogr.*, **55** (1983) 136.
53. S. Chakravorty and C. M. Wayman, *Metall. Trans.*, **7A** (1976) 555.
54. Y. Murakami, K. Otsuka, S. Hanada and S. Watanabe, *Mater. Sci. Eng.*, **A189** (1994) 191.
55. L. E. Tanner, D. Schryvers and S. M. Shapiro, *Mater. Sci. Eng.*, **A127** (1990) 205.
56. K. Enami, V. V. Martynov, T. Tomie, L. G. Khandros and S. Nenno, *Trans. JIM*, **22** (1981) 357.
57. Y. Kudoh, M. Tokonami, S. Miyazaki and K. Otsuka, *Acta Metall.*, **33** (1985) 2049.
58. T. Hara, T. Ohba and K. Otsuka, *Suppl. J. de Phys.*, III, (Proc. ICOMAT-95) **5** (1995) C8-641.
59. O. Matsumoto, S. Miyazaki, K. Otsuka and H. Tamura, *Acta Metall.*, **35** (1987) 2137.
60. S. Miyazaki, S. Kimura, K. Otsuka and Y. Suzuki, *Scr. Metall.*, **18** (1984) 883.
61. T. Saburi, M. Yoshida and S. Nenno, *Scr. Metall.*, **18** (1984) 363.
62. S. Miyazaki, S. Kimura and K. Otsuka, *Phil. Mag.*, **57A** (1988) 467.
63. T. Saburi, K. Doi and S. Nenno, *Mater. Sci. Forum*, (Proc. ICOMAT-89), **56–58** (1990)

611.

64. C. M. Hwang, M. Meichle, M. Salamon and C. M. Wayman, *Phil. Mag.*, **47A** (1983) 9, 31.
65. Y. Shugo, F. Hasegawa and T. Honma, *Bull. Res. Inst. Mineral Dressing and Metallurgy* (Tohoku Univ.), **37** (1981) 79.
66. T. H. Nam, T. Saburi and K. Shimizu, *Mater. Trans. JIM*, **31** (1990) 959.
67. K. Enami, Y. Kitano and K. Horii, *Proc. MRS Int. Meeting on Adv. Mater.*, **9** (Tokyo, 1988) 117.
68. P. G. Lindquist and C. M. Wayman, *Proc. MRS Int. Meeting on Adv. Mater.*, **9** (Tokyo, 1988) 123.
69. V. N. Khachin, N. M. Matveeva, V. I. Sivokha and D. B. Chernov, *Dokl. Akad. Nauk USSR*, **257** (1981) 1676.
70. K. Otsuka, K. Oda, Y. Ueno and Min Piao, *Scr. Metall.*, **29** (1993) 1355.
71. D. Golberg, Ya Xu, Y. Murakami, S. Morito, K. Otsuka, T. Ueki and H. Horikawa, *Scr. Metall.*, **30** (1994) 1349.
72. L. Guttman, *Trans. AIME*, **188** (1950) 1472.
73. M. W. Burkart and T. A. Read, *Trans. AIME*, **197** (1953) 1516.
74. Z. S. Basinski and J. W. Christian, *Acta Metall.*, **2** (1954) 101.
75. S. Miura, M. Ito, K. Endo and N. Nakanishi, *Memoirs Faculty Eng.*, Kyoto Univ., **18** (1981) 287.
76. Y. Koyama and S. Nittono, *J. Jpn. Inst. Met.*, **43** (1979) 262. [in Japanese]
77. F. T. Worrell, *J. Appl. Phys.*, **19** (1948) 929.
78. E. Z. Vintaikin, D. F. Litvin, V. A. Udovenko and G. V. Scherbedinskij, *Proc. Int. Conf. on Martensitic Transformations (ICOMAT-79)*, (Cambridge, 1979) 673.
79. T. Saburi and S. Nenno, *Scr. Metall.*, **8** (1974) 1363.
80. T. Takezawa and S. Sato, Proc. 1st JIM Int. Symp. on New Aspects of Martensitic Transformations, *Suppl. Trans. JIM*, **17** (1976) 233.
81. T. A. Schroeder and C. M. Wayman, *Scr. Metall.*, **11** (1977) 225.
82. R. Oshima and E. Naya, *J. Jpn. Inst. Met.*, **39** (1975) 175. [in Japanese]
83. M. Nishida and T. Honma, *Bull. Res. Inst. Mineral Dressing and Metallurgy* (Tohoku Univ.), **37** (1981) 79. [in Japanese]
84. A. Ölander, *J. Am. Chem. Soc.*, **56** (1932) 3819.
85. S. Miura, S. Maeda and N. Nakanishi, *Phil. Mag.*, **30** (1974) 565.
86. H. Sakamoto, K. Otsuka and K. Shimizu, *Scr. Metall.*, **11** (1977) 607.
87. G. Barcelo, R. Rapacioli and M. Ahlers, *Scr. Metall.*, **12** (1978) 1069.
88. Y. Nakajima, S. Aoki, K. Otsuka and T. Ohba, *Mater. Let.*, **21** (1994) 271.
89. M. W. Burkart and T. A. Read, *Trans. AIME*, **197** (1953) 1516.
90. Z. S. Basinski and J. W. Christian, *Acta Metall.*, **2** (1954) 101.
91. H. K. Birnbaum and T. A. Read, *Trans. AIME*, **218** (1960) 662.
92. K. Otsuka and C. M. Wayman, in *Deformation Behavior of Materials*, ed. P. Feltham, Vol. **II** (Freund Publishing House, Israel, 1977) 151.
93. T. Ohba, K. Otsuka and S. Sasaki, *Mater. Sci. Forum*, **56–58** (1990) 317.
94. K. Marukawa and K. Tsuchiya, *Scr. Metall. et Mater.*, **32** (1995) 77.
95. T. Suzuki, T. Tonokawa and T. Ohba, *Suppl. J. de Phys.*, **III** (ICOMAT-95), **5** (1995) C8-1065.
96. H. Funakubo (ed.), *Shape Memory Alloys*, Gordon Breach Sci. Pub., NY, (1987).
97. H. Suzuki (ed.), *Proc. 1st JIM Int. Symp. on New Aspects of Martensitic Transformations (ICOMAT-76)*, (Jpn. Inst. Met., Sendai, 1976).
98. W. S. Owen *et al.* (ed.), *Proc. Int. Conf. on Martensitic Transformations (ICOMAT-79)*, (Cambridge, Massachusetts, 1979).
99. L. Delaey and S. Chandrasekaran, *J. de Phys.*, **43**, Suppl. 12, (Proc. ICOMAT-82), (1982).
100. I. Tamura (ed.), *Proc. Int. Conf. on Martensitic Transformations (ICOMAT-86)*, (Jpn. Inst. Met., Sendai, 1987).
101. B. C. Muddle (ed.), Martensitic Transformations (Proc. ICOMAT-89), *Mater. Sci. Forum*, **56–58** (1990).
102. C. M. Wayman and J. Perkins (ed.), *Proc. Int. Conf. on Martensitic Transformations*

(*ICOMAT-92*), (Monterey Institute of Advanced Studies, Carmel, CA, 1993).

103. R. Gotthardt and J. Van Humbeeck, Proc. Int. Conf. on Martensitic Transformations (ICOMAT-95), *Suppl. J. de Phys.*, **III**, No. 12 (1995).
104. K. Otsuka and K. Shimizu (ed.), *Proc. MRS Int. Meeting on Adv. Mater.* **9** (Tokyo, 1988), (MRS, Pittsburgh, 1989).
105. C. T. Liu, H. Kunsmann, K. Otsuka and M. Wuttig (ed.), Shape Memory Materials – Fundamental Aspects and Applications, *MRS Symp. Proc.* **246** (1992).
106. K. Otsuka and Y. Fukai (ed.), *Shape Memory Materials and Hydrides, Advanced Materials '93/B*, **18B** (Elsevier, Amsterdam, 1994).
107. E. P. George, S. Takahashi, S. T. McKinstry, K. Uchino and M. Wun-Fogle (ed.), Materials for Smart Systems, *MRS Symp. Proc.*, **360** (1995).
108. E. P. George, R. Gotthardt, K. Otsuka, S. Troiler-McKinstry and M. Wun-Fogle (ed.), Materials for Smart Systems II, *MRS Symp. Proc.*, **459** (1997).
109. J. Perkins (ed.), *Shape Memory Effect in Alloys*, (Plenum Press, NY, 1975).
110. T. W. Duerig, K. N. Melton, D. Stockel and C. M. Wayman (ed.), *Engineering Aspects of Shape Memory Alloys* (Butterworth–Heinemann, London, 1990).
111. E. Hornbogen and N. Jost (ed.), *The Martensitic Transformation in Science and Technology*, (Informazionsgesellschaft, Oberursel, 1989).
112. Chu Youyi, T. Y. Hsu and T. Ko (ed.), *Proc. Inst. Symp. on Shape Memory Alloys* (Guilin, 1986), (China Academic Pub., 1986).
113. Chu Youyi and Tu Hailing (ed.), *Proc. Int. Symp. on Shape Memory Materials* (Beijing, 1994), (International Academic Pub., Beijing, 1994).
114. A. R. Pelton, D. Hodgson and T. Duerig (ed.), *Proc. 1st Int. Conf. on Shape Memory and Superelastic Technologies* (NDC, Inc., Fremont, California, 1995).
115. X. Ren and K. Otsuka, *Nature*, **389**, No. 6651 (1997) 579.

3

Ti–Ni shape memory alloys

T. SABURI

Symbols

A_f finish temperature of the B19′ → B2 (or R) transformation
$A_{f'}$ finish temperature of the R → B2 transformation
A_s start temperature of the B19′ → B2 (or R) transformation
$A_{s'}$ start temperature of the R → B2 transformation
M_f finish temperature of the B2 (or R) → B19′ transformation
M_s start temperature of the B2 (or R) → B19′ transformation
R_f finish temperature of the R-phase transformation
R_s start temperature of the R-phase transformation
X_{Ni} total nickel content
Y_{Ni} nickel content corrected for the oxide Ti_4Ni_2O
X_O total oxygen content
Y_M stress of yielding due to stress-inducing of the B19′ martensite from the R-phase
Y_R stress of yielding due to rearrangement of the R-phase variants
$\varepsilon_{B19'}$ strain associated with the R → B19′ transformation
ε_R strain associated with the B2 → R transformation
η strain recovery ratio

3.1 Structure and transformations

In near-equiatomic Ti–Ni alloys, the shape memory effect[1] and transformation pseudoelasticity[2] occur in association with the thermoelastic martensitic transformation from the parent phase (β) with a B2 structure to the phase with a monoclinic B19′ structure, or more often in association with the two-step transformation from the β to a trigonal phase (so called R-phase) and then to the B19′ phase. The phenomena are sensitive to the fine structure of the parent β phase. Therefore, factors such as Ni content, aging, thermo-mechanical treatment and addition of alloying elements, which affect the structure, are important for controlling the memory behavior.

3.1.1 Partial phase diagram of near-equiatomic compositions

Many investigations have been made on the equilibrium phase diagram of the Ti–Ni system[3–9] and the B2 phase region is known to be very narrow at temperatures below 923 K (650 °C).[5,7,9] Figure 3.1 is the equilibrium Ti–Ni

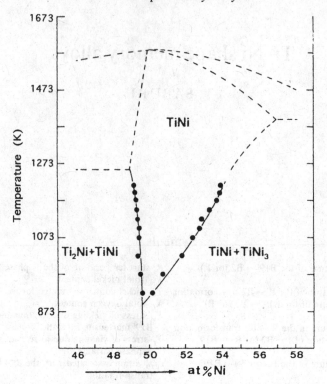

Fig. 3.1. Equilibrium phase diagram for Ti–Ni alloys in the vicinity of TiNi. (After Bastin and Rieck[9])

phase diagram in the vicinity of TiNi, which has been proposed by Bastin and Rieck.[9] Although, in this phase diagram, the B2 phase region below 923 K is not specifically shown, it is generally accepted that the B2 region is only between 50.0 and 50.5 at% Ni.

3.1.2 Decomposition of near-equiatomic B2 phase of Ti–Ni alloys

Ti–Ni alloys with nickel contents exceeding 50.5 at% decompose on cooling slowly from a high temperature or on aging at a temperature below 973 K (700°C) after quenching from a high temperature. For example, a 52 at% Ni alloy decomposes in the way shown by the time–temperature–transformation (TTT) curves in Fig. 3.2,[10] when aged at various temperatures after quenching from 1273 K (1000°C). As seen in Fig. 3.2, there are the following three temperature ranges, in each of which the decomposition scheme is unique.

(a) aging at temperatures below 953 K (680°C)

$\quad$ TiNi → TiNi + Ti_3Ni_4 → TiNi + Ti_2Ni_3 → TiNi + $TiNi_3$

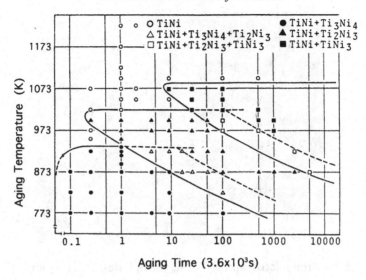

Fig. 3.2. Time–temperature–transformation curve of a Ti–52 at% Ni alloy. (After Nishida *et al.*[10])

(b) aging at temperatures between 953 K and 1023 K (750°C)
  $TiNi \rightarrow TiNi + Ti_2Ni_3 \rightarrow TiNi + TiNi_3$
(c) aging at temperatures between 1023 K and 1073 K (800°C)
  $TiNi \rightarrow TiNi + TiNi_3$

In any of the three temperature ranges, the final product of decomposition is a mixture of $TiNi_3$ and $TiNi$. The Ti_3Ni_4 and Ti_2Ni_3 phases are metastable. In the range (a), the three phases Ti_3Ni_4, Ti_2Ni_3 and $TiNi_3$ appear in this sequence, which is the order of increasing nickel content of the product phases. In the range (b), only the Ti_2Ni_3 phase appears before the $TiNi_3$ appears, and in the range (c) the $TiNi_3$ forms directly from the initial $TiNi$.

The Ti_3Ni_4 phase forms in the early stages of aging at low temperatures as fine platelets with coherency to the matrix (see Fig. 3.3[11]) and affects the properties of the TiNi alloys. The Ti_3Ni_4 phase has a rhombohedral structure as will be discussed in Section 3.1.3.

The Ti_2Ni_3 phase which has been interpreted to be a product of a peritectoid reaction $(TiNi + TiNi_3 \rightarrow Ti_2Ni_3)$[7] is now understood to form as an intermediate phase prior to the formation of the final product, $TiNi_3$. The eutectoid reaction $(TiNi \rightarrow Ti_2Ni + TiNi_3)$ which was once proposed[3,6,8] does not exist. It appears that Ti_4Ni_2O was misinterpreted as Ti_2Ni since these two phases are similar in structure. The Ti_4Ni_2O oxide forms easily and is commonly observed in Ti–Ni alloys.

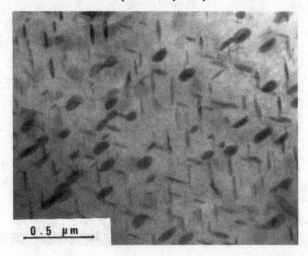

0.5 μm

Fig. 3.3. Electron micrograph of Ti_3Ni_4 precipitates. (After Saburi[11])

3.1.3 Crystal structure of Ti_3Ni_4

Ti_3Ni_4 has a rhombohedral structure which belongs to the space group R3.[12–14] Six titanium atoms and eight nickel atoms compose the rhombohedral unit cell. The lattice parameters in the rhombohedral system are $a = 0.6704$ nm, $\alpha = 113.85$.

When we use the hexagonal axis system which is more convenient to describe the structure, the atom positions in a unit cell are: (0, 0, 0; 1/3, 2/3, 2/3; 2/3, 1/3, 1/3) + Ti: (18f) + (x, y, z; y, x − y, z; y − x, x, z) with $x = 5/7$, $y = 1/7$, $z = 0$: Ni (1): (18f) + (x, y, z; y, x − y, z; y − x, x, z) with $x = 4/21$, $y = 5/21$, $z = 1/6$; Ni(2): (3b) 0, 0, 1/2; Ni (3): (3a) 0, 0, 0.

The atom positions in the hexagonal unit cell are illustrated in Fig. 3.4.[12] The unit cell is composed of the six layers A, B, C, D, E and F of Fig. 3.4(a). There are 18 titanium atoms and 24 nickel atoms in the unit cell and the ratio between them is $18:24 = 3:4 (= 43:57)$. This structure is derived from the parent B2 structure by placing Ni atoms on the open circle positions in the A, C and E layers which were originally occupied by Ti atoms in the parent B2 structure (entire sites of the A, C and E layers are occupied by Ti in the parent state). The Ti_3Ni_4 phase had been called X-phase before the structure was made clear. The Ti–Ni ratio of the X-phase which was experimentally obtained by analytical electron microscopy[10,13] was 44:56. This is very close to the ratio 43:57 mentioned above.

The orientation relationship between Ti_3Ni_4 (referred to the hexagonal axes) and the matrix (M) is,

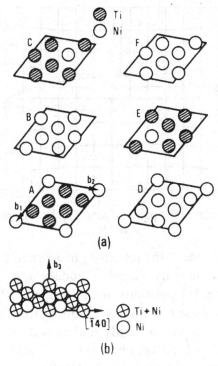

(a)

(b)

Fig. 3.4. Structure of Ti_3Ni_4. (After Saburi *et al.*[12])

$$(0001)_X \parallel (111)_M$$
$$[010]_X \parallel [2\bar{1}3]_M,$$

where X is Ti_3Ni_4 and M is matrix. The precipitate particles of Ti_3Ni_4 are lenticular in shape and the habit plane is $(111)_M$ (or $(0001)_X$). Ti_3Ni_4 shrinks 2.7% along the $[001]_X$ direction and 0.3% along its perpendicular directions (in the hexagonal system) relative to the matrix. The precipitates in the early stage are coherent to the matrix and produce strain fields around them. Figure 3.5 illustrates the situation.[14] These strain fields are important in producing the all-round shape memory effect.[15] Precipitation of Ti_3Ni_4 strengthens the matrix B2 phase and thus improves the recoverability of the shape memory effect.[16,17]

3.1.4 Martensitic transformations in Ti–Ni alloys

While fully annealed near-equiatomic Ti–Ni alloys transform from the B2 parent phase directly to the monoclinic B19′ phase martensitically, thermally cycled or thermo-mechanically treated near-equiatomic Ti–Ni alloys transform in two steps, i.e., from the B2 parent to the R-phase and then to the B19′

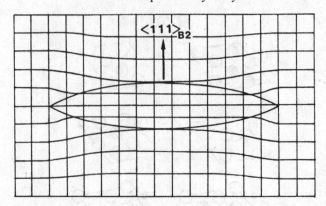

Fig. 3.5. Lattice distortion of the parent due to Ti_3Ni_4 precipitate. (After Tadaki
et al.[14])

phase.[18–21] The appearance of the resistivity increase in the resistivity versus
temperature curve after thermal cycling[22,23] and the appearance of additional
X-ray reflections at the 1/3 positions in the reciprocal lattice[22] before the
appearance of the B19′ phase reflections were once proposed to be premarten-
sitic phenomena, but now are generally understood as due to the B2 → R
transformation. The B2 → R transformation is also martensitic. The two-step
transformation also occurs in nickel-rich Ti–Ni alloys aged at an appropriate
temperature[17] and in ternary Ti–Ni–Fe[24–30] and Ti–Ni–Al[31] alloys. Figure
3.6[32] shows the transformation sequence of a thermo-mechanically treated
49.8 Ti–50.2 Ni (at%) alloy. When the specimen is cooled from 373 K, electrical
resistivity starts to increase at R_s on the electrical resistivity versus temperature
curve (Fig. 3.6(a)). This temperature coincides with the start temperature of the
first peak in the differential calorimetry (DSC) curve (R_s in Fig. 3.6(b)). The
start of the resistivity increase and that of the DSC peak coincide within 2 K.[33]
This coincidence indicates that the resistivity increase and the first DSC peak
are due to the R-phase transformation which occurs as a first order transformation, and that R_s is the start temperature of the R-phase transformation. The
second peak in the DSC curve is due to the transformation from the R-phase to
the B19′ phase. The temperature at which electrical resistivity starts to decrease coincides with the start temperature of the second peak in the DSC
curve. This is the M_s temperature, i.e., the temperature at which the R → B19′
transformation starts.

   The B2 → R transformation takes place by nucleation and growth. Figure
3.7[34] demonstrates the nucleation and growth process of the R-phase in a
Ti–48Ni–2Al (at%) alloy. Figure 3.7(a) is an electron micrograph of the parent
phase (note the dislocations marked 1 and 2). On cooling, thin plates of
R-phase nucleate from the dislocations 1 and 2 (Fig. 3.7(b)), grow and join

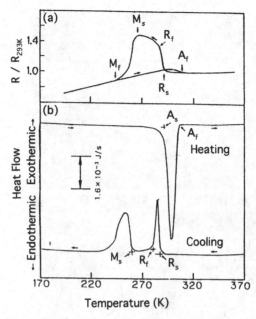

Fig. 3.6. Electrical resistivity versus temperature curve (a) and DSC curve (b) of a thermo-mechanically treated 49.8Ti–50.2Ni (at%) alloy. (After Nam *et al.*[32])

together (Fig. 3.7(c)). On further cooling, many other plates of the R-phase form similarly, and finally the entire region changes into the R-phase. On heating, the R-phase plates shrink and disappear. The appearance and disappearance repeat in the same manner on repeated cooling and heating.

The temperature at which electrical resistivity starts to increase was once interpreted as the temperature where a second order transformation to an incommensurate phase starts.[26] However, it is now understood that the incommensurate phase does not exist and the resistivity increase is due to the R-phase transformation.

Figure 3.8(a) is an electron diffraction pattern of the parent phase (above R_s) of the Ti–48Ni–2Al (at%) alloy and Fig. 3.8(b) is the corresponding bright field image.[35] It is to be noted that there are diffuse diffraction spots in addition to the strong B2 lattice spots. The diffuse spots are located at near 1/3 positions along $\langle 110 \rangle$ of the B2 reciprocal lattice. The near 1/3 position spots are very weak and their dark field images reveal no specific structure. Let us call the corresponding phase B2' in order to distinguish it from the B2 phase which does not give rise to the diffuse 1/3 spots. On cooling below R_s, transformation takes place from the B2', which gives rise to the diffuse 1/3 spots, to the R-phase which gives rise to the sharp 1/3 spots at the exact positions. Figure 3.8(c) is the electron diffraction pattern of the same area as that of Fig. 3.8(a)(b) at a temperature below R_s and Fig. 3.8(d) is the corresponding bright field image.

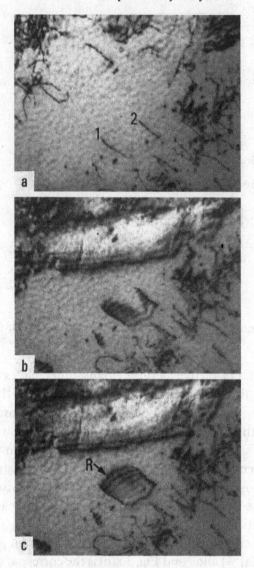

Fig. 3.7. Electron micrographs showing R-phase nucleation from single dislocations (Ti–48Ni–2Al (at%) alloy). (After Fukuda et al.[34])

The 1/3 position spots are sharp and lie on the exact positions. The bright field image shows the typical self-accommodation structure of the R-phase variants, which will be discussed in Section 3.4.1. The diffuse near-1/3 reflections are observed at temperatures of a relatively wide range above R_s.[22] The origin of the diffuse near-1/3 reflections has not been made clear yet. The B2 → B2′ transformation may be a soft mode transformation.[29,36] The B2′ → R-phase

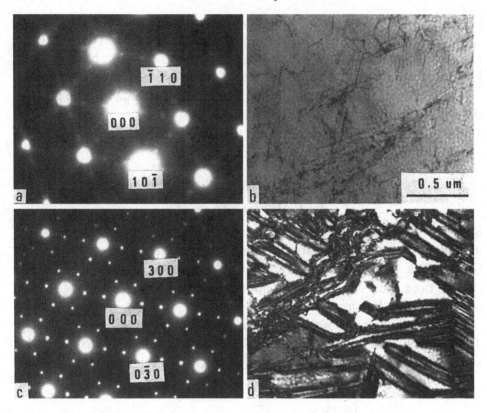

Fig. 3.8. Electron diffraction patterns of parent (a) and of R-phase (c), and the corresponding bright field images of the parent (b) and of the R-phase (d), Ti–48Ni–2Al (at%). (After Saburi[35])

transformation is one of the typical first order transformations which take place heterogeneously by nucleation and growth.

3.1.5 Crystal structures of the B19'- and the R-phases

The B19' martensite has a monoclinic structure[37–41] and it belongs to the space group $P2_1/m$.[38–41] Lattice parameters (a, b, c and β) for a Ti–49.2 at% Ni alloy are: $a = 0.2898$ nm, $b = 0.4108$ nm, $c = 0.4646$ nm and $\beta = 97.78$ and they are composition dependent.[41] The structure is different from the 2H type stacking structure of martensites in Cu-based alloys in that the shufflings during the transformation are not parallel to the basal plane in the former.[41] The unit cell of the B19' phase is elongated about 10% relative to the parent along the parent $[223]_{B2}$ orientation and this defines the maximum recoverable shape change along this direction.[42–44]

The R-phase has a trigonal structure which is described by a hexagonal lattice for convenience sake. The lattice parameters are $a_R = 0.738$ nm and $c_R = 0.532$ nm.[36] Although the trigonal structure of the R-phase was once reported to belong to the space group $P\bar{3}_1m$,[36] Hara *et al.*[45] have recently found that it belongs to the space group P3. The R-phase is elongated 0.94% along the parent [111] orientation.[21,30] This is one order smaller than that for the B19′ martensite.

3.2 Mechanical behavior of Ti–Ni alloys

3.2.1 Stress–strain curve

Stress–strain curves of the near-equiatomic Ti–Ni alloys are characterized by discontinuous yielding and large Lüders strains, when tested at temperatures between 77 K (-196°C) and 152 K (75°C).[46,47] At a temperature above 373 K (100°C) the Lüders strain does not appear and more than 15% elongation is observed before fracture.

Figure 3.9[35] is the tensile stress–strain curve of a wire specimen of a Ti–50 at% Ni alloy (diameter: 0.8 mm). The specimen was annealed for 3.6 ks at 673 K after cold-drawing and tested at 303 K which is below R_f (316 K) and

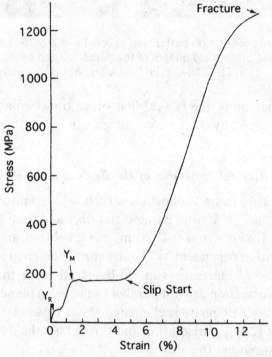

Fig. 3.9. Stress–strain curve of a Ti–50.0 at% Ni alloy wire. (After Saburi[35])

above M_s (246 K) and thus the specimen was entirely in the R-phase state when tested. On stressing, the first yielding occurs at Y_R which is the starting point of the deformation due to rearrangement of the R-phase variants. This first step due to the variant rearrangement gives rise to an elongation of about 0.8%. The second yielding occurs at Y_M which is the starting point of the deformation due to stress-induction of the B19′ martensite from the R-phase. This second step due to the stress-induced B19′ martensite gives rise to an elongation of about 5% including that of the R-phase. After this, the rate of stress-increase becomes larger and slip deformation begins. Finally, fracture occurs at about 15% elongation. If the test temperature is below M_f, the first step does not occur, since no R-phase exists. In this case, only one large step which is due to rearrangement of the variants of the B19′ martensite occurs. If the test temperature is above R_s, only one step due to the stress-induced B19′ appears before slip deformation starts.

3.2.2 Shape memory effect and pseudoelasticity: effect of thermo-mechanical treatment

As was mentioned in Section 3.1.1, the single phase region of the B2 ordered high temperature phase in the phase diagram of Ti–Ni is very narrow (from 50.0 at% to 50.5 at% Ni) at temperatures below 923 K (650 °C). While properties of the alloys of Ni contents exceeding 50.5 at% are sensitive to heat treatment at temperatures between 573 K (300 °C) and 773 K (500 °C) due to the resulting precipitation of Ti_3Ni_4, those of the alloys of Ni contents between 50.0 and 50.5 at% are insensitive to heat treatment because no precipitation of the Ti_3Ni_4 occurs in these alloys. However, thermo-mechanical treatments (annealing at a temperature below 773 K after cold-working) affect their properties very much.

Near-equiatomic Ti–Ni alloys (of Ni-content less than approximately 50.5 at% Ni), when fully annealed, behave pseudoelastically only partially.[16,17] The series of stress–strain curves in Fig. 3.10[16] illustrate the mechanical behavior of a Ti–50.2 at% Ni alloy fully annealed at 1123 K (850 °C). Temperatures in the figure refer to the test temperature. In the tests, tensile stress was applied till about 4% elongation was reached and then the stress was removed. It is seen in the figure that the yield stress becomes minimum at about 282 K (9 °C) which coincides with the temperature at which the martensitic transformation starts on cooling; i.e., $M_s = 282$ K. In the tests at temperatures below 293 K (20 °C), the residual elongation after unloading vanishes completely on heating above A_f, as shown by the arrows in the figure. This is the shape memory effect.

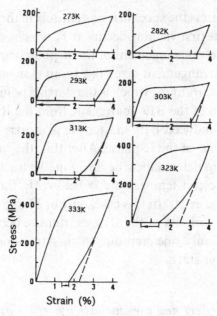

Fig. 3.10..Stress–strain curves of a Ti–50.2 at% Ni alloy annealed at 1123 K (850°C) for 3.6 ks and water-quenched. (After Saburi *et al.*[16])

In shape memory alloys generally, transformation pseudoelasticity should appear in a certain temperature range above A_f. However, in the present case, pseudoelastic shape recovery takes place only partially. The deviation from the usual elastic recovery shown by the broken lines for the stress–strain curves above 293 K (20°C) in Fig. 3.10 is partial pseudoelasticity. For the tests above 313 K (40°C) the residual elongation after unloading vanishes only partially on heating above A_f; the shape memory is incomplete. This permanent elongation is due to slip deformation and it increases with increasing test temperature. In this way, for the fully annealed near-equiatomic Ti–Ni alloys, pseudoelasticity is incomplete at any temperature above A_f and shape memory also is not good. However, this can be improved by thermo-mechanical treatment which effectively increases the resistance for slip deformation and prevents plastic deformation by slip;[17] thus, the pseudoelastic strain increases substantially. Figure 3.11[48] shows the effect of thermo-mechanical treatment. There, the specimens were cold-rolled as much as 6 to 25% and then annealed at 673 K (400°C) for 3.6 ks. Degree of cold-rolling is shown above each stress–strain curve. The tensile tests were made at 323 K (50°C), which is about 40 K above the M_s. It is clearly seen in the figure that the strain recoverable on unloading increases with increasing degree of cold-rolling prior to annealing. When the alloy is cold-rolled more than 20% and then annealed at 673 K (400°C), it shows complete pseudoelasticity. Figure 3.12[48] shows a series of stress–strain curves

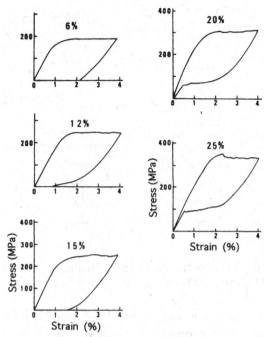

Fig. 3.11. Improvement of pseudoelasticity achieved by cold-rolling followed by annealing at 313 K (400 °C). Degree of cold-rolling is shown above each stress–strain curve. Tensile tests were made at 323 K (50 °C). (After Saburi et al. [48])

of a Ti–50.2 at% Ni alloy annealed at 673 K for 3.6 ks after cold-rolling 25% and tested at various temperatures. At temperatures below 303 K (30 °C), pseudoelastic strain recovery does not occur, although strain recovery on heating after unloading (shape memory effect) does occur as shown by arrows in Fig. 3.12. At 313 K (40 °C), pseudoelasticity appears, and above 323 K pseudoelastic recovery is 100%. As seen at bottom right of Fig. 3.12, tensile strain of as much as 7% vanishes completely on unloading.

The recrystallization temperature of the near-equiatomic Ti–Ni alloys, when cold-rolled 25% and annealed, lies between 773 K (500 °C) and 873 K (600 °C) as seen in Fig. 3.13; in specimens annealed at a temperature below 773 K, apparent deformation structures exist, whereas in the specimen annealed at 873 K, recrystallized small grains appear and the recrystallized grains become larger as annealing temperature is raised. Specimens cold-rolled 25% and annealed at 873 K, which is nearly the recrystallization temperature, behave like those annealed at 1123 K (850 °C) (see Fig. 3.10) and pseudoelastic strain recovery is only partial. This indicates that annealing in the thermo-mechanical treatment has to be done at an appropriate temperature below the recrystallization temperature for the high dislocation-density introduced by cold-working to be maintained.

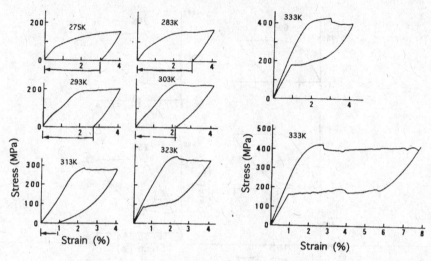

Fig. 3.12. Stress–strain curves of the Ti–50.2 at% Ni alloy annealed at 313 K (400 °C) for 3.6 ks after 25% cold-rolling. (After Saburi *et al.*[48])

As nickel content exceeds 50.5 at%, the alloys tend to behave pseudoelastically at temperatures in a certain range above M_s, even in a fully annealed and water-quenched condition, and the temperature range of pseudoelasticity in a fully annealed and quenched condition shifts towards the low temperature side with increasing nickel content.[16,17] The temperature range of pseudoelasticity is between 333 K (-60 °C) and 303 K (-30 °C) for a 51.3 at% Ni alloy, and on aging at a temperature below 773 K (500 °C) it shifts to the high temperature side due to precipitation of Ti_3Ni_4 particles.[16] This is understood to be due to enrichment of titanium in the matrix which results from the precipitation of Ti_3Ni_4.

3.2.3 *Orientation dependence of shape memory characteristics and effect of texture*

The shape memory strain (the strain which is recoverable by shape memory effect) originates from the characteristic shape-change which results from the lattice deformation of the associated transformation. Accordingly, the maximum strain recoverable for a single crystal can be predicted theoretically.[42] The predicted values for the B2 (cubic) → B19′ (monoclinic) transformation in the equiatomic Ti–Ni are as follows. The recoverable elongation is largest along $[233]_{B2}$ and it is 10.7%. It is 9.8% along $[111]_{B2}$, 8.4% along $[011]_{B2}$ and 2.7% along $[001]_{B2}$. The recoverable contraction is 4.2% along $[001]_{B2}$, 5.2% along $[011]_{B2}$ and 3.6% along $[111]_{B2}$. These predicted values have been confirmed by tensile experiments using single crystal specimens.[43,44]

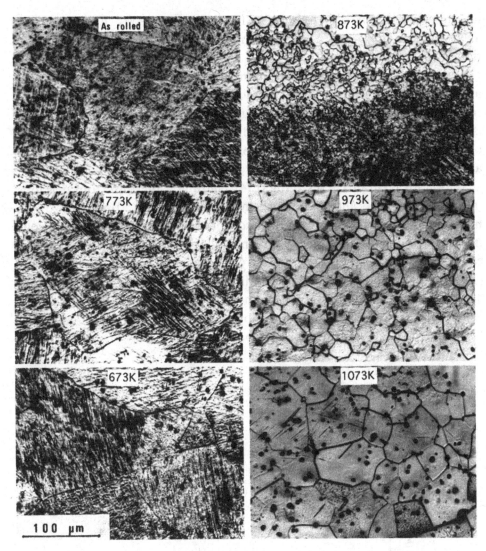

Fig. 3.13. Optical micrographs showing recrystallization in a Ti–50.2 at% Ni alloy. Annealed for 3.6 ks at each temperature after 25% cold-rolling. Annealing temperature is shown at the top of each photograph. (After Saburi *et al.*[48])

Since we use shape memory alloys for practical purposes in a polycrystalline state, it is important to know about the anisotropy of the polycrystalline aggregates. While a polycrystal shape memory material in which crystal grains are oriented at random should be isotropic in its shape memory behavior, a material with a preferred orientation should be anisotropic. Our knowledge about the recoverable strain of a TiNi single crystal mentioned above is useful in assessing the effect of the preferred orientation. If we can develop a suitable texture, we should be able to improve the memory capacity. For example, a TiNi wire having a $[111]_{B2}$ fiber axis should give a large recoverable elongation in the length direction.

A hot-rolled and annealed sheet (1 mm thick) of TiNi has a texture described as (112) $[110]_{B2}$ (result by X-ray diffraction), and the sheet gives the largest recoverable elongation along the rolling direction.[49] This experimental result is consistent with the prediction above that $[110]_{B2}$ is one of the favorable orientations for recoverable elongation; predicted recoverable elongation along $[110]_{B2}$ is 8.4%.

Tensile stress–strain curves in Fig. 3.14[11] show how the anisotropy develops in a Ti–50.8 at% Ni alloy. There, an ingot of $30 \times 50 \times 50$ mm in size was hot-rolled, and specimens were prepared from several different stages of rolling: as-cast ingot, 10 mm (thickness), 5 mm, 2.4 mm, 1.5 mm and 0.8 mm. For each stage, tensile test specimens were cut parallel (R.D.) and perpendicular (T.D.) to the rolling direction. The tensile tests were done on these specimens after annealing at 1073 K (800°C) and then aging at 773 K (500°C) for 3.6 ks. It is clearly seen in the figure that the alloy is isotropic in the as-cast condition and anisotropy develops as rolling proceeds.

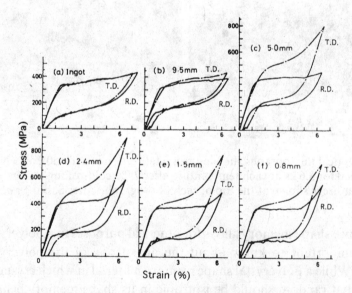

Fig. 3.14. Development of anisotropy in a Ti–50.8 at% Ni alloy by hot-rolling. Stress–strain surves for specimens prepared from several different stages of hot-rolling are shown. R.D.; rolling direction. T.D.; transverse direction. Test temperature; 303 K (30°C) (20 K above A_f). (After Saburi[11])

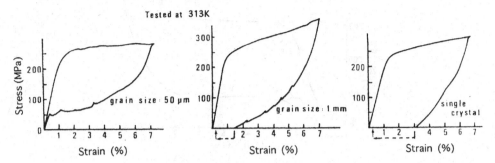

Fig. 3.15. Grain size dependence of pseudoelasticity in Ti–50.5 at% Ni alloy. (After Saburi *et al.*[43])

3.2.4 *Grain size dependence of mechanical behavior*

While single crystal specimens of a near-equiatomic Ti–Ni alloy show incomplete pseudoelasticity at temperatures above A_f, polycrystal specimens of the same alloy show a complete pseudoelasticity at temperatures above A_f. Figure 3.15[43] illustrates the grain size dependence of recoverability of a Ti–50.5 at% Ni alloy which was annealed at 1073 K (800°C) for 3.6 ks and quenched. The stress–strain curves in the figure were obtained by tensile-testing the specimens with three different grain sizes: (a) 50 μm, (b) 1 mm and (c) single crystal. The tests were made at 313 K (40°C) which is 10 K above the A_f (= 303 K (30°C)). It is clearly seen in the figure that the reduction of grain size is very effective for improving the pseudoelasticity.

3.2.5 *Shape memory effect and pseudoelasticity associated with the R-phase*

R-phase transformation occurs preceding the transformation to the B19′ martensite in thermo-mechanically treated near-equiatomic Ti–Ni alloys,[17-21] in nickel rich Ti–Ni alloys aged at an appropriate low temperature,[16,17] and also in ternary Ti–Ni–Fe[24-30] and Ti–Ni–Al[31] alloys. The R-phase transformation is martensitic and thermoelastic. Thus, this R-phase transformation also gives rise to shape memory effect and pseudoelasticity in association with it.[17-21,30,48,50-54]

In contrast to the B2 → B19′ transformation which yields a large shape change (approximately 7%), the B2 → R transformation yields only a small shape change (approximately 0.8%) and therefore the recoverable strain associated with this transformation is small. However, the temperature hysteresis of this transformation is also small. For example, for a Ti–Ni coil spring, the hysteresis of a temperature versus shear strain curve under a constant applied stress is about 1.5 K,[53] which is very much smaller than the hysteresis of the

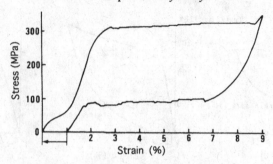

Fig. 3.16. Stress–strain curve of a Ti–47.5 Ni–2.5 Fe (at%) alloy sheet annealed at 773 K (500 °C) for 1 hour after cold-rolling 30%; stressed till the end of the second stage and unloaded; tested at 193 K (-80 °C) (below R_s and above A_s). (After Saburi et al.[48])

$B2 \rightarrow B19'$ transformation which is greater than 40 K. This small hysteresis is good for applications as sensors.

When a thermo-mechanically treated near-equiatomic Ti–Ni alloy or Ti–Ni–Fe alloy is tested at a temperature between R_s and M_s by a conventional tensile testing method, two steps appear successively in the stress–strain curve before slip begins; the first step is small (0.8% elongation) and the second is large (7% elongation).[21,30] Figure 3.16[48] is an example of the two-step stress–strain curve obtained with a Ti–47.5Ni–2.5Fe (at%) alloy annealed at 773 K (500 °C) after cold-rolling. The test was performed at 193 K (-80 °C) (48 K below R_s and 30 K above M_s). The first step is associated with reorientation of the R-phase variants and the second step with stress-induction of the B19′ martensite. Since the test temperature is between R_s and M_s, a complete pseudoelasticity associated with the stress-induced $R \rightarrow B19'$ transformation occurs. The residual strain is due to the rearrangement of the R-phase variants. This strain vanishes on heating above R_s.

In many practical applications of the shape memory alloys, they are used in such a way that they yield an abrupt shape change at a specific temperature during thermal cycling under load. Therefore it is very important to know the deformation behavior under this type of conditions. For this purpose a thermal cycling test under a constant load is useful. Figure 3.17[52,53] is an example of a strain–temperature curve under constant load (102 MPa) for a wire specimen of Ti–49.8 at% Ni alloy annealed at 743 K after cold-drawing. The specimen was cooled to a temperature below M_f. It is seen in the figure that two steps appear on cooling. The first step at R_s is due to the appearance of R-phase and the second at M_s is due to the appearance of B19′ martensite under the applied stress (102 MPa). Only one large step appears at about 355 K on heating. The large step is due to the reverse transformation from the B19′ martensite directly to the B2 parent.

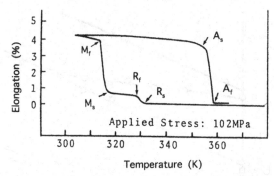

Fig. 3.17. Strain–temperature curve for B2 → R → B19′ transformation of a Ti–49.8 at% Ni wire annealed at 743 K (470°C) after cold-drawing. Applied stress: 102 MPa. (After Todoroki and Tamura[52])

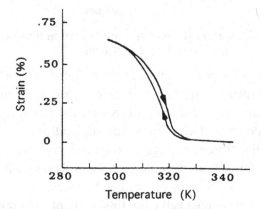

Fig. 3.18. Strain–temperature curve for B2 → R transformation of a Ti–50.2 at% Ni wire annealed at 773 K for 900 s after cold-drawing. Applied stress: 200 MPa. (After Stachowiak and McCormick[54])

When the specimen is cooled to a temperature below R_s but above M_s under load and then heated, a reverse step due to the R → B2 transformation appears with a small hysteresis of 2 K. Figure 3.18[54] shows an example for a specimen heat-treated at 773 K and tested under a stress of 200 MPa. If the same specimen is cooled to a temperature between M_s and M_f under the same stress and then heated, it exhibits a two-step strain recovery on heating. The first step is associated with the R → B2 transformation and the second step with the B19′ → B2 transformation.[54]

Figure 3.19[54] shows the effect of applied stress on the transformation temperature. All the transformation temperatures increase linearly with increasing applied stress. R_s exhibits a stress dependence substantially smaller than that of M_s or A_f. For stresses greater than 355 MPa, M_s exceeds R_s and only one step (due to the direct formation of B19′ from B2) appears in the temperature elongation curve on cooling.

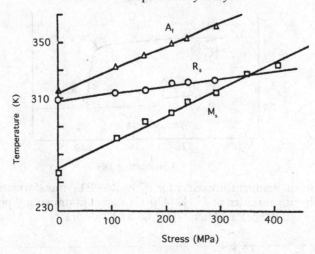

Fig. 3.19. The effect of applied stress on transformation temperatures. (After
Stachowiak and McCormick[54])

For applications in which a reversible shape change with a small hysteresis
at a specific temperature is required, such as for sensor-actuator purposes, the
B2 → R transformation is useful and in this case the transformations B2 → R
and R → B19′ should be well separated. For applications in which large scale
shape recovery with sufficient strength is required, such as for pipe coupling,
the B2 → B19′ transformation (including the B2 → R) is useful. In this case, the
large hysteresis is not a problem but rather beneficial.

Todoroki et al.[50–53] extensively did this type of constant-load thermal
cycling tests on Ti–Ni alloys thermo-mechanically treated in various ways, in
their course developing an actuator device for an air-conditioner. They made
tests both on straight wire specimens (diameter: 0.75 mm) and on coil specimens (wire diameter: 0.75 mm, coil diameter: 5.6 mm). By plotting the critical
temperatures inducing the R-phase and the B19′ martensite against the applied
stresses, they obtained a phase relation similar to the one in Fig. 3.19. The data
obtained by constant-load thermal cycling tests (transformation temperatures
under external stress, sizes of the shape changes, sizes of the hysteresis) are
useful for designing devices using shape memory alloys.

3.2.6 Effects of thermal cycling and deformation cycling

The effect of thermal cycling through the B19′ transformation under no load is
shown in Fig. 3.20(a)(b).[55] M_s (start temperature of the B2 → B19′ or R → B19′
transformation) decreases with increasing number of cycles (N), but R_s (start
temperature of the B2 → R) is not affected. Thus, in alloys where M_s initially
exceeds or is nearly equal to R_s, the decrease in M_s caused by thermal cycling

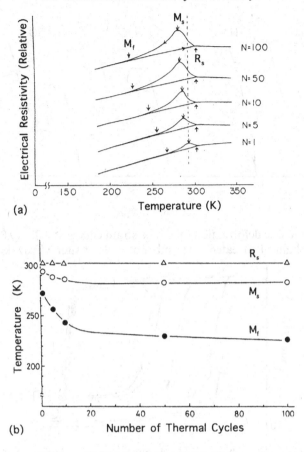

Fig. 3.20. (a) Effect of thermal cycling on electrical resistivity versus temperature curve of a Ti–49.8 at% Ni alloy which was annealed at 1273 K for 3.6 ks. The arrows indicate R_s, M_s and M_f. N: number of cycles. (b) Transformation temperatures plotted against number of thermal cycles. (After Miyazaki *et al.*[55])

results in the situation where R_s is higher than M_s. With further increase in the number of thermal cycles, the separation between R_s and M_s increases. This leads to an increasing amount of R-phase and enhancement of the resistivity peak due to the B2 → R transformation as seen in Fig. 3.20(a). This decrease in M_s occurs due to introduction of dislocations during the repeated motion of the parent–martensite interface and is inevitable for well-annealed (and quenched) Ti–Ni alloys. However, this can be avoided by thermo-mechanical treatment (for near stoichiometric alloys) or by aging (for alloys with nickel exceeding 50.5 at%).[55]

By deformation cycling in the temperature range of pseudoelasticity (above A_f), the shape of the pseudoelasticity loop in the stress–strain curve changes gradually; yield stress and hysteresis width of the loop decrease and permanent

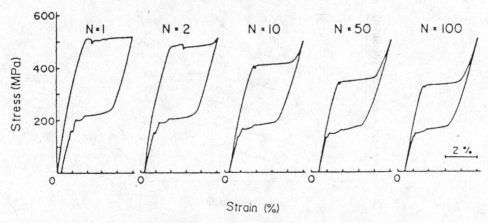

Fig. 3.21. Effect of cyclic deformation on stress–strain curves of a Ti–49.8 at% Ni alloy
thermo-mechanically treated. N: number of cycles. (After Miyazaki et al.[56])

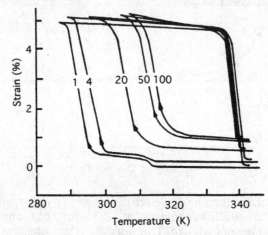

Fig. 3.22. Influence of repeated thermal cycling on transformation behavior of
a Ti–50.2 at% Ni wire annealed at 773 K for 900 s after cold-drawing. Applied
stress: 200 MPa (numbers refer to the number of cycles). (After Stachowiak and
McCormick[54])

elongation increases (see Fig. 3.21).[56] Although the change in shape of the
pseudoelasticity loop is significant in early cycles, it becomes very little after
100 cycles. Thus, it has been suggested that training by deformation cycling
prior to actual service is effective for stabilizing the pseudoelasticity.[56]

Figure 3.22[54] shows the effect of thermal cycling under load on the trans-
formation strain of the B2 → R (ε_R) and that of the R → B19′ ($\varepsilon_{B19'}$). Both ε_R and
$\varepsilon_{B19'}$ decrease with increasing number of cycles. After 20 cycles, ε_R vanishes.
Thermal cycling under load, in a temperature range where B19′ martensite is
induced, raises M_s and reduces the temperature hysteresis between the forward
and reverse transformations (but has little effect on A_f).[53,54] The increase in M_s

with increasing number of cycles, causing overlap of the B2 → R and B2 → B19′ transformations, is opposite to the result of thermal cycling under no load.[54] This M_s increase is due to the dislocations produced during thermal cycling. The dislocation structure developed during thermal cycling assists the preferred nucleation of the variants suitable for the applied stress during cooling, thus raising the M_s.[54]

Thermal cycling under load, in a temperature range such that the R-phase is repeatedly induced and the B19′ martensite is not, does not cause any appreciable change in the pseudoelasticity loop even after 500 000 cycles, as is shown in Fig. 3.23.[53] It appears that the B2 → R transformation cycling induces little substructural change.

3.2.7 Effects of oxygen and carbon

Figure 3.24(a)[57] shows the M_s temperatures plotted against total (soluble and insoluble) oxygen contents (X_O (in at%)) in some specially prepared Ti–Ni–O alloys. The measurements were made on specimens quenched from 973 K into water. It is obvious that M_s decreases linearly with increasing oxygen content. Since solid solubility of oxygen in the Ti–Ni alloys is very small (0.045 at%), alloys of oxygen contents exceeding this limit solidify, on cooling from a melt, into primary TiNi and a eutectic mixture of a solid solution (TiNi) and an oxide (Ti_4Ni_2O). The composition of the oxide has been confirmed by electron probe X-ray micro-analysis (EPMA). In this way, most of the oxygen atoms in

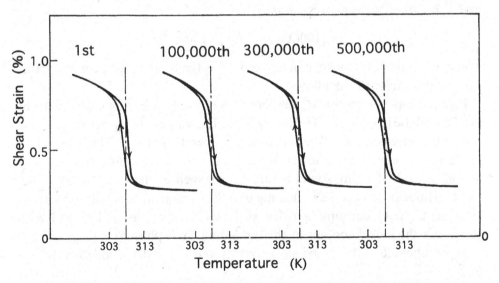

Fig. 3.23. Effect of thermal cycling under load on the R-phase transformation of a Ti–49.8 at% Ni coil annealed at 748 K (475 °C) after cold-drawing. (After Todoroki[53])

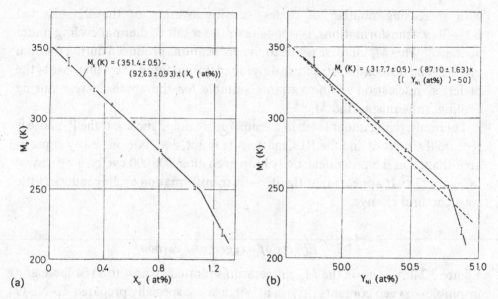

Fig. 3.24. (a) M_s temperatures of Ti–Ni–O alloys plotted against total oxygen
contents (X_O (at%)). (b) M_s temperatures of Ti–Ni–O alloys plotted against nickel
contents in TiNi phase (Y_{Ni}) corrected for Ti_4Ni_2O. (After Shugo et al.[57])

Ti–Ni alloys in the solid state exist in the form of Ti_4Ni_2O, thus consuming
twice as much Ti than Ni in the remaining TiNi B2 phase. If one assumes that
oxygen is entirely insoluble in the TiNi phase and all the oxygen atoms exist
forming Ti_4Ni_2O oxide, the nickel content in the TiNi phase corrected for the
oxide, Y_{Ni} (in at%) is expressed as

$$Y_{Ni} = \{100(X_{Ni} - 2X_O)\}/\{100 - 7X_O\}$$

where X_{Ni} is the total nickel content and X_O is the total oxygen content (both
expressed in at%) in the alloy.

Figure 3.24(b) shows the M_s temperatures plotted against nickel contents in
the TiNi phase (Y_{Ni}) in the Ti–Ni–O alloys. The values of Y_{Ni} were calculated
using the above equation. The solid line is the least-square fit. The broken line
in the figure shows the relation between M_s and nickel content experimentally
obtained for Ti–Ni alloys with a very low oxygen content ($X_O = 0.09$ at%).
The two lines are very close, indicating that the oxygen atoms really exist in the
form of Ti_4Ni_2O consuming twice as much titanium as nickel, and thus
increasing the nickel content in the remaining TiNi phase.

Ti_4Ni_2O oxide in the Ti–Ni alloys hardly affects the shape memory behavior.[57]

From M_s temperature measurements on some specially prepared Ti–Ni–C
alloy specimens (quenched from 1028 K into water), Shugo et al.[58] found that

M_s decreases with increasing carbon content. When the M_s temperatures of the Ti–Ni–C alloys are plotted against nickel contents in the TiNi phase, which are calculated assuming that all the carbon atoms exist as TiC, a linear relation is obtained and the line is parallel to, but located approximately 15 K below, the line for carbon free Ti–Ni alloys. This indicates that carbon really precipitates as TiC and consumes titanium in the matrix and also that a small amount of carbon is soluble and the dissolved carbon in the TiNi matrix lowers the M_s temperature by about 15 K.[58]

Carbides do not affect the yield stress and the yield strain, but fracture strain tends to be increased by carbon addition to 0.2–0.5 at%.[58]

3.3 Ternary alloying

3.3.1 *Effects of ternary alloying addition on transformation temperature*

The shape recovery temperature is highest at the stoichiometric composition (50 at% Ti–50 at% Ni) in the binary Ti–Ni system. It is about 393 K (120 °C). It decreases with increasing nickel content and does not change with decreasing nickel content.

Ternary alloying additions affect the transformation temperatures, and thus the shape recovery temperatures of Ti–Ni alloys. Substitution of vanadium, chromium, manganese or aluminum for titanium lowers the transformation temperatures.[59] Substitution of cobalt or iron for nickel also lowers the martensitic transformation temperatures.[60] Substitution of iron for nickel separates the transformation temperature ranges of the $B2 \rightarrow R$ and the $R \rightarrow B19'$ effectively.[25,26] At present, palladium[61] and gold[60] are the only alloying elements which are known to be effective for raising the transformation temperatures. There have been controversial reports for zirconium substitution for titanium. While Eckelmeyer[60] reports that zirconium substitution raises the recovery temperature, Honma *et al.*[59] report that it lowers the M_s temperature.

No one of the simple factors, such as atomic size, ionic size, electron negativity and electron–atom ratio can be correlated to the influence of the ternary additions on the shape recovery temperature of Ti–Ni alloys.[60]

Substitution of copper for nickel is discussed separately in Section 3.3.2.

3.3.2 *Ti–Ni–Cu alloys*

Ternary Ti–Ni–Cu alloys have been investigated extensively in view of improving the shape memory and pseudoelasticity characteristics.[32,62–71] Figure

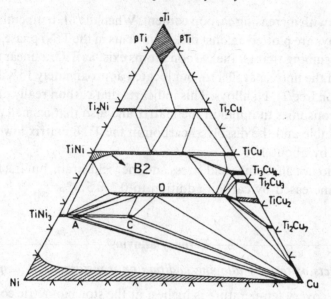

Fig. 3.25. Ti–Ni–Cu phase diagram at 1143 K (870°C). (After Vanloo *et al.*[72])

3.25 is the equilibrium Ti–Ni–Cu phase diagram at 1143 K.[72] It is seen in the diagram that in Ti–Ni–Cu alloys in which Cu is substituted for Ni, the parent phase remains as a single B2 phase up to 30 at% Cu. These Ti–Ni–Cu alloys show shape memory effect.[63] However, substitution of Cu for Ni affects the transformation behavior and the associated shape memory characteristics in many respects.[32,62–71]

Ti–Ni–Cu alloys of Cu contents from 5 to 15 at% (Cu is substituted for Ni) transform in two steps, i.e., from cubic to orthorhombic (B19), and then to monoclinic B19′. With increasing Cu content, the transformation start temperature of the B2 → B19 does not change, but that of the B19 → B19′ decreases and in alloys of copper contents exceeding 15 at%, the second transformation does not occur even at liquid nitrogen temperature.[63,67]

The transformation hysteresis of the B2 → B19 in the Ti–Ni–Cu alloys is much smaller than that of the B2 → B19′ transformation in the binary alloys.[63,66] This small hysteresis is closely related to the ease of interface movement during the transformation.[64] Copper addition also affects the mechanical behavior of Ti–Ni alloys in many respects. Figure 3.26[66] shows the elongation–temperature curves of Ti–Ni–Cu alloys obtained by constant load thermal cycling tests. It is seen that with increasing Cu content the transformation hysteresis is very much reduced. While the hysteresis of the B2 → B19 transformation in a binary Ti–Ni alloy is more than 40 K (as seen in Fig. 3.17), that of the Ti–Ni–10 Cu (at%) alloy is 11 K and the recoverable strain is more

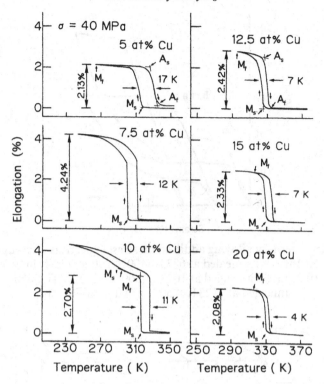

Fig. 3.26. Elongation–temperature curves of Ti–Ni–Cu alloys obtained by thermal cycling under a constant load. (After Nam et al.[66])

than 3%. Thus the B2 → B19 transformation is useful for applications as actuators. Unfortunately, Cu addition exceeding 10 at% embrittles the alloy and spoils the formability.

Copper addition reduces the stress-hysteresis in the pseudoelasticity (difference in stress level between during loading and during unloading in stress–strain curve).[11,65] Figure 3.27[11] illustrates the effect. It is seen in the figure that the hysteresis of the Ti–40.5 Ni–10.0 Cu alloy is less than 100 MPa, while that of the binary alloy is more than 200 MPa and that of the Ti–Ni–Nb alloy is about 300 MPa.

Figure 3.28[65] shows the effect of Cu addition on flow-stress. The flow-stress level in the martensitic state is reduced by Cu addition. Compare the curves in Fig. 3.28 with those in Fig. 3.12 which are for a binary Ti–Ni alloy. Rearrangement of martensite variants appears to be much easier in the Ti–Ni–Cu alloys.

Cu addition is also effective for avoiding aging effects; it prevents Ti_3Ni_4 precipitation and thus it is effective for avoiding M_s temperature change due to differences in cooling speed. Composition sensitivity of M_s is also very much reduced by Cu addition.[65]

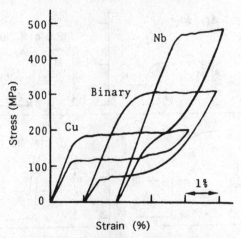

Fig. 3.27. Effect of ternary alloying addition on stress-hysteresis of pseudoelasticity. Binary: Ti–50.2 at% Ni annealed at 673 K (400°C) after cold-rolling 30%. Cu: Ti–40.0Ni–10.0Cu (at%) annealed at 1073 K (800°C). Nb: 48Ti–50Ni–2Nb (at%) annealed at 1123 K (850°C). (After Saburi[11])

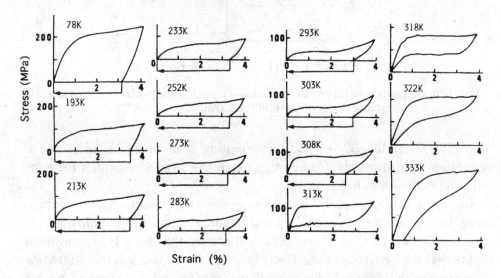

Fig. 3.28. Tensile stress–strain curves of the 49.5Ti–40.5Ni–10.0Cu (at%) alloy quenched from 1123 K. Temperatures in the figure refer to test temperatures. (After Saburi *et al.*[65])

3.3.3 *Ti–Ni–Nb alloys*

A wide transformation hysteresis in shape memory alloys is convenient for coupling devices.[73] The transformation temperatures are adjusted so that ambient temperature is located near the middle of the wide hysteresis ranging from liquid nitrogen temperature to a temperature higher than ambient temperature. Expansion of a device is done at liquid nitrogen temperature and the expanded state can be kept at ambient temperature. Thus the coupling device can be shipped and stored at ambient temperature in the martensitic state, i.e., in the as-expanded shape. Installation of the device is done by heating it to an appropriate temperature. Thus liquid nitrogen cooling is not needed. The ternary addition element known to be effective for hysteresis widening is Nb (typically 44Ti–47Ni–9Nb (at%)).[74]

The as-cast 44Ti–47Ni–9Nb (at%) alloy consists of primary TiNi phase surrounded by a eutectic structure of the TiNi and a nearly-pure Nb phase. The primary TiNi contains a small amount of Nb and the nearly-pure Nb contains small amounts of Ni and Ti.[74,75] The microstructure of the alloy after hot-working is characterized by dispersion of fine particles of the nearly-pure Nb phase in the TiNi matrix.[76,77] The Nb particles are very soft, with a flow stress similar to that of the B19′ martensite of TiNi (a flow stress of 150–200 MPa). Thus, substantial plastic deformation occurs in the Nb particles, during deformation of the martensite by reorientation of variants. This partitioning of the strain into a reversible part (the TiNi matrix) and an irreversible part (Nb particles) is the reason for the large thermal hysteresis of the Ti–Ni–Nb alloys.[77–80]

Predeformation of more than 12% (overdeformation) in the martensitic state is needed for maximizing the hysteresis of the Ti–Ni–Nb alloys. The effect of the overdeformation is to increase A_s for the first heating cycle. From the second cycle, it reverts to its original value. M_s and M_f are not affected by the overdeformation.[74,78,79]

The relation between the hysteresis width defined by $(A_s - M_s)$ and the total strain, and the relation between the strain recovery ratio (η) and the total strain are shown in Fig. 3.29(a).[78] It is clear that when the total strain exceeds 16%, the transformation hysteresis jumps from 40 K to 145 K and the recovery ratio is still sufficiently large at 16% of strain.

The relation between the transformation hysteresis and the deformation temperature, and the relation between the recovery ratio and the deformation temperature are shown in Fig. 3.29(b).[78] It is seen that as the deformation temperature reaches M_s (188 K ($-85°$C)) from the lower temperatures, the transformation hysteresis increases rapidly from 40 K to 145 K and the recovery ratio remains sufficiently large.

3.3.4 Ti–Pd–Ni alloys

Near-equiatomic Ti–Pd alloys have a B2 structure at temperatures above about 783 K (510°C), and it transforms into a B19 structure below this tem-

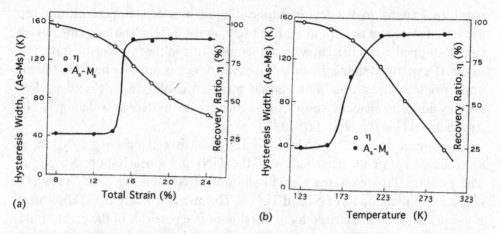

(a)

(b)

Fig. 3.29. (a) Effect of deformation on the transformation hysteresis $(A_s - M_s)$ and strain recovery ratio (η). (b) Effect of deformation temperature on the transformation hysteresis $(A_s - M_s)$ and strain recovery ratio (η). (After Zhang *et al.*[78])

perature.[81,82] The single phase region of the TiPd B2 phase below 1000 K (727°C) is very narrow (from 49.0 at% Pd to 50.0 at% Pd).[83] The transformation from B2 to B19 is martensitic.[84,85]

TiNi and TiPd form pseudobinary solid solutions (B2 order phase) at high temperatures and the pseudobinary alloys transform martensitically; alloys of the TiPd side transform from cubic (B2) to orthorhombic (B19) and those of the TiNi side transform from cubic (B2) to monoclinic (B19′).[61] Figure 3.30[61] shows the composition dependence of the transformation temperatures of the pseudobinary alloys. The M_s temperature can be varied from ambient temperature to 783 K by controlling the alloy composition. Thus Ti–Ni–Pd alloys are promising for use in high temperature shape memory devices.

However, high transformation temperatures do not guarantee good shape memory characteristics at high temperatures.[86] The critical stress for slip deformation generally decreases with increasing temperature and slip deformation easily occurs concurrently or preceding the deformation by reorientation of the martensite variants on stressing the alloy at a high temperature. Therefore, fully annealed TiPd and Ti–Pd–Ni alloys exhibit only a partial shape memory.[86,87]

Thermo-mechanical treatment, consisting of cold-rolling and subsequent annealing at an appropriate temperature below recrystallization temperature, is effective for improving the high temperature shape memory characteristics.[87] Figure 3.31[87] shows the effect of thermo-mechanical treatment. The stress–strain curves in Fig. 3.31 are the results of tensile tests on a $Ti_{50}Pd_{30}Ni_{20}$ alloy at high temperatures. The curves in Fig. 3.31(a) are the

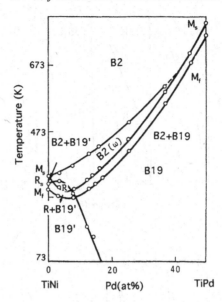

Fig. 3.30. M_s temperature of pseudobinary TiNi–TiPd alloys. R_s: transformation temperature of R-phase. B2(ω): B2 phase with an ω-like distortion. (After Matveeva *et al.*[61])

results on the fully-annealed specimens and those in (b) are the results on the thermo-mechanically treated (annealing at 673 K for 3.6 ks after cold-rolling 24.5%) specimens. The arrows in the figure indicate the recovery of strain on heating above A_s after unloading (the shape memory strain). It is clear that permanent residual strain at high temperatures below 520 K vanishes in the thermo-mechanically treated specimens. Interestingly, the thermo-mechanically treated specimen exhibits pseudoelastic behavior at 535 K, which is 45 K above A_s and 11 K below A_f.

When palladium of the TiPd alloy is substituted by Fe (or Cr), the transformation temperatures decrease with increasing Fe (or Cr) content.[88,89] Thus, Fe (or Cr) addition is convenient for controlling the transformation temperatures. The shape memory effect appears in the TiPd–Fe and TiNi–Cr alloys.[89] In the TiPd–Fe and TiNi–Cr alloys, martensites of 2H and 9R structures and, in addition, an intermediate phase (incommensurate phase) appear.[88,89]

3.4 Self-accommodation in martensites

3.4.1 R-phase

Between the R-phase and the B2 parent phase, there are four equivalent lattice correspondences shown in Table 3.1.[34] The resulting four orientation variants

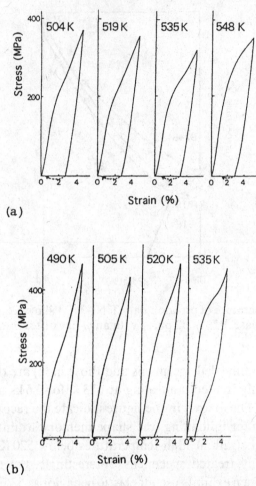

Fig. 3.31. Tensile stress–strain curves of fully-annealed (a) and thermo-mechanically treated (b) $Ti_{50}Pd_{30}Ni_{20}$ alloys. Temperatures in the figure refer to test temperatures. (After Golberg et al.[87])

of the R-phase are twin related to each other and there are two types of twin relations. One is the $(11\bar{2}1)_H [11\bar{2}]_H$ type compound twin and the other is the $(11\bar{2}\bar{2})_H [111]_H$ type compound twin. Any two of the four variants are twin related by either one of these two schemes. Self-accommodation of the R-phase is achieved by arranging these twin-related four variants around one of the $\langle 001 \rangle_{B2}$ axes as is shown in Fig. 3.32,[34] where the parent indices are used for convenience instead of the hexagonal indices of the R-phase. The variants 1 and 3 (and 2 and 4) in Fig. 3.32 are $(11\bar{2}\bar{2})_H$ twins and the variants 2 and 3 (and 1 and 4) are $(11\bar{2}1)_H$ twins. The $(11\bar{2}\bar{2})_H$ is derived from one of the parent $\{110\}_{B2}$ and the $(11\bar{2}1)_H$ is derived from one of the parent $\{100\}_{B2}$. Six kinds of equivalent arrangements are possible (two for each of the $\langle 001 \rangle_{B2}$ axes of the

Table 3.1. *Lattice correspondence between the B2 parent and the triclinic R-phase. B2: B2 lattice of the parent. H: hexagonal lattice of the R-phase.* (*After Fukuda* et al.[34])

Variant	$[100]_H$	$[010]_H$	$[001]_H$	$(0001)_H$
1	$[1\bar{2}1]_{B2}$	$[11\bar{2}]_{B2}$	$[111]_{B2}$	$(111)_{B2}$
2	$[211]_{B2}$	$[\bar{1}1\bar{2}]_{B2}$	$[\bar{1}11]_{B2}$	$(\bar{1}11)_{B2}$
3	$[121]_{B2}$	$[112]_{B2}$	$[\bar{1}\bar{1}1]_{B2}$	$(\bar{1}\bar{1}1)_{B2}$
4	$[211]_{B2}$	$[11\bar{2}]_{B2}$	$[1\bar{1}1]_{B2}$	$(1\bar{1}1)_{B2}$

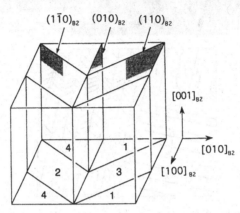

Fig. 3.32. Model of self-accommodation in the R-phase. (After Fukuda *et al.*[34])

parent) among the four variants relative to the parent lattice. A typical example of the self-accommodation of the R-phase is shown in Fig. 3.33(a) (an optical micrograph of a Ti–48.2 Ni–1.3 Fe (at%) alloy). There are two types of morphology. One is straight-band morphology and the other is herring-bone (saw-tooth) morphology. The former corresponds to the view from $[100]_{B2}$ or $[010]_{B2}$ and the latter to the view from $[001]_{B2}$ of Fig. 3.32. The variant numbers assigned on the basis of the lattice correspondence in Table 3.1 are shown in Fig. 3.33(b).

3.4.2 B19 martensite

Between the orthorhombic B19 martensite and the B2 parent there are six equivalent lattice correspondences shown in Table 3.2.[90] The resulting six orientation variants of the B19 are twin-related and there are two types of twin relations also in this case. One is the $(111)_O$ twin (type I twin) and the other is the $(011)_O$ twin (compound twin). Any two of the six variants are twin-related by either one of these two schemes. The basic self-accommodation scheme is

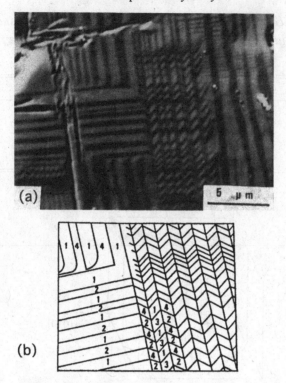

Fig. 3.33. Optical micrograph of self-accommodation of the R-phase
(Ti–48.2Ni–1.5Fe (at%) alloy). (After Fukuda *et al.*[34])

shown in Fig. 3.34(a).[90] The scheme is characterized by six variant units, each
of which is a pentahedron bounded by four $(111)_O$ type twins $((111)_O$ is derived
from one of the parent $\{110\}_{B2}$), and a $(011)_O$ type twin $((011)_O$ is derived from
one of the parent $\{001\}_{B2}$). The variants 2 and 4 are in a $(111)_O$ twin relation,
and the variants 3 and 4 are in a $(011)_O$ twin relation, for example. Triangular
morphology should appear when the structure in Fig. 3.34(a) is observed in the
$\langle 111 \rangle_{B2}$ direction, as is shown in Fig. 3.34(b). Figure 3.35[90] is an electron
micrograph of boundary traces of variants in the B19 martensite of a
Ti–40.5 Ni–10.0 Cu (at%) alloy. The specimen normal is parallell to the
$[111]_{B2}$ and the morphology is characterized by triangles as expected from
Fig. 3.34(b).

3.4.3 B19′ martensite

Between the monoclinic B19′ martensite and the B2 parent there are twelve
lattice correspondences. Self-accommodation is characterized by triangular
morphology[91,92] but is much more complicated than the B19 martensite.

Table 3.2. *Lattice correspondence between the B2 parent and the orthombic B19 martnesite. B2: B2 lattice of the parent. O: orthombic lattice of the B19. (After Watanabe et al.[90])*

Variant	$[100]_O$	$[010]_O$	$[001]_O$
1	$[100]_{B2}$	$[011]_{B2}$	$[0\bar{1}1]_{B2}$
2	$[100]_{B2}$	$[0\bar{1}1]_{B2}$	$[0\bar{1}\bar{1}]_{B2}$
3	$[010]_{B2}$	$[101]_{B2}$	$[10\bar{1}]_{B2}$
4	$[010]_{B2}$	$[10\bar{1}]_{B2}$	$[\bar{1}0\bar{1}]_{B2}$
5	$[001]_{B2}$	$[110]_{B2}$	$[\bar{1}10]_{B2}$
6	$[00\bar{1}]_{B2}$	$[1\bar{1}0]_{B2}$	$[\bar{1}\bar{1}0]_{B2}$

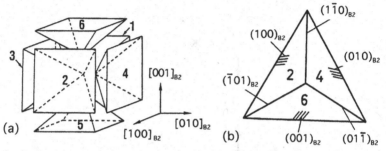

Fig. 3.34. (a) Model of self-accommodation in the B19 martensite. (b) View from $[111]_{B2}$ orientation of the model in (a). (After Watanabe *et al.*[90])

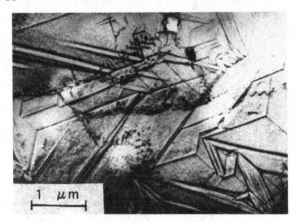

Fig. 3.35. Electron micrograph of self-accommodation of the B19 martensite of a Ti–40.5Ni–10.0Cu (at%) alloy. (After Watanabe *et al.*[90])

During the transformation from the B2 phase or from the R phase, transformation twins as the lattice invariant shear occur in each of the martensite plates. The transformation twins are $\langle 011 \rangle$ type II twins[91,92] and these transformation twins make the self-accommodation scheme complicated.

3.5 All-round shape memory (Two-way shape memory)

Specially treated Ni-rich Ti–Ni alloys exhibit a reversible shape memory phenomenon called all-round shape memory, which belongs to a category of the two-way shape memory. An example of the observed all-round shape memory effect is shown in Fig. 3.36.[93] The specimen in the figure is a 0.3 mm thick sheet of a Ti–51.0 at% Ni alloy (Fig. 3.36(a)). This is heat-treated under

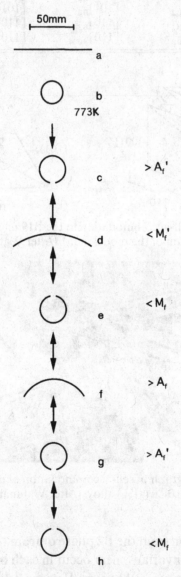

Fig. 3.36. All-round shape memory of a 49Ti–51Ni (at%) alloy. (After Nishida and Honma[93])

constraint in the following way. It is forced to fit along the inner circle of a copper tube (diameter of 20 mm) used as a jig (as shown in Fig. 3.36(b)), thus being deformed into a round-shape. Then it, still in the jig, is heat-treated at 773 K to imprint the round shape. When the as heat-treated specimen is removed from the jig, it takes a round-shape of a diameter slightly larger than that of the jig, if the temperature is above the R → B2 transformation-finish temperature A_f' which is near room temperature, as is shown in Fig. 3.36(c) where the temperature is 373 K (that of boiling water). When this is taken out from the boiling water and cooled in air, its shape changes spontaneously into that of Fig. 3.36(d). This spontaneous shape change between (c) and (d) is associated with the B2 → R transformation and thus the temperature hysteresis is very small. When the specimen is cooled further, the specimen flattens and rounds upward. The shape at a temperature below M_f (213 K) is shown in Fig. 3.36(e). It is seen that the specimen is in an upward round-shape, which is opposite to the downward round-shape at 373 K in Fig. 3.36(c). The shape change from (d) to (e) is associated with the R → B19' transformation and the temperature hysteresis is large compared to that of the B2 → R. The shape change from (c) to (e) is reversible. The upward curvature below M_f shown in Fig. 3.36(e) reverts to the downward curvature when the specimen is heated above A_f (Fig. 3.36(f)) and returns to the original downward round-shape completely when heated to above A_f' (Fig. 3.36(g)). When cooled below M_f again the specimen takes the upward round-shape again (Fig. 3.36(h)).

In the all-round shape memory, Ti_3Ni_4 precipitates play an important role. The mechanism of the effect proposed by Kainuma *et al.*[15] is as follows.

As already mentioned in Section 3.1.3, Ti_3Ni_4 particles are lenticular in shape and the habit plane is $(111)_M$ $((001)_X)$. The particles are coherent to the matrix in the early stages of aging and shrink 2.7% along the $[001]_X$ direction. Therefore, when precipitation takes place under a specific externally-applied stress, the precipitate particles align themselves; under a tensile stress, the habit plane, $(001)_X$, of the particles tends to be parallel to the tensile stress, and under a compressive stress, it tends to be perpendicular to the compressive stress, as shown in Fig. 3.37(a) and (b). When the external stress is removed, the particles create a tensile stress perpendicular to their habit plane, due to their shrinkage relative to the matrix.

When a sheet specimen is deformed into a round-shape as shown in Fig. 3.37(a), the outer part should be in a tensile-stress state and the inner part in a compressive-stress state. Therefore, when the specimen is aged in this state, the habit plane of the particles tends to be parallel to the specimen surface in the outer part and perpendicular to the surface in the inner part (Fig. 3.37(c)). On cooling from this state, the two-step transformation B2 → R → B19', takes

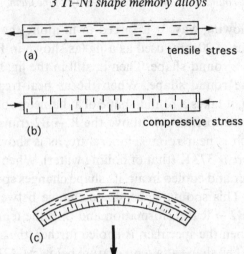

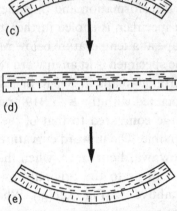

Fig. 3.37. Mechanism of the all-round shape memory. (After Kainuma et al.[15])

place under the control of the tensile-stress fields produced by the precipitate particles. Thus the variants of the R-phase (and B19′ martensite later), which are effective to cancel the tensile stress, form preferentially. In the outer part, the preferential variants are of the kind which yield the largest expansion perpendicular to the surface (the largest contraction parallel the surface), while in the inner part, the preferential variants are of the kind which yields the largest expansion parallel to the surface (the largest contraction perpendicular to the surface). Thus the total shape of the specimen changes as shown in Fig. 3.37(d) and (e), i.e., from the downward round-shape to the upward round-shape.

For the all-round shape memory to occur, precipitation of Ti_3Ni_4 is necessary, and thus the Ni content of the alloy has to be greater than 50.6 at% and the aging temperature has to be above 673 K. However, coherency strain produced by the Ti_3Ni_4 precipitates is also important and the particle size has to be small (typically 150 nm). Therefore, aging temperature should not exceed

823 K, above which the particle size becomes too large and the coherency strain is lost.

3.6 Effects of irradiation on the shape memory behavior

Coupling devices using Ti–Ni alloys are attracting interest for applications in the field of nuclear engineering. The shape memory couplings are thought to be convenient for remote-control operation in replacing structural components of reactors. However, for this purpose, we must know the effects of irradiation on the shape memory behavior.

Neutron irradiation at temperatures near ambient temperature suppresses the martensitic transformations in Ti–Ni alloys.[94–97] However, annealing at temperatures above 523 K after the irradiation restores the original condition. Neutron irradiation at temperatures above 423 K induces little change in the transformation behavior.[98]

An observed example of the effect on the mechanical behavior of Ti–50.3 at% Ni alloy is shown in Fig. 3.38.[95] The specimens were fully annealed initially. They were then irradiated by neutrons in a reactor. The irradiation was done at 323 K with a fluence of $3 \times 10^{23}\,\mathrm{m}^{-2}$ of fast neutrons. Then tensile tests were done on these specimens at various temperatures between 173 K and 373 K. The obtained stress–strain curves of the irradiated specimens are shown by solid lines. The stress–strain curves obtained with the specimens prior to irradiation are shown by the broken lines. The mechanical behavior after irradiation is unique. Tensile strain of 5% vanishes pseudoelastically with no hysteresis on unloading. This behavior appears to be different from the usual transformation pseudoelasticity. Nevertheless, it is also different from the ordinary elasticity; the modulus is too small. It appears that both the stress-induced martensitic transformation and the slip deformation do not occur in the irradiated specimens at test temperatures between 173 K and 373 K. The original mechanical property is restored by annealing at temperatures above 550 K.

The changes in the transformation behavior and in the mechanical properties are thought to be due to the changes in local lattice structure.[97]

3.7 Sputter-deposited films of Ti–Ni alloys

As a promising method for producing micro-actuators, sputter deposition of films of Ti–Ni alloys is attracting interest.[99–114] Pre-alloyed targets are usually used to obtain thin films of a desired composition. However, films sputter-deposited from the alloy target show Ti depletion and Ni enrichment, prob-

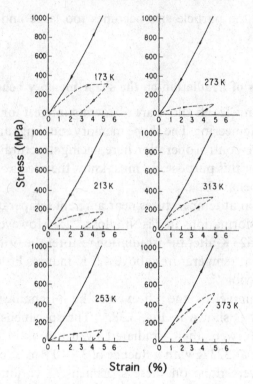

Fig. 3.38. Stress–strain curves of a neutron-irradiated Ti–50.3 at% Ni alloy at various temperatures. Temperatures in the figure refer to test temperatures. (After Hoshiya *et al.*[95])

ably due to oxidation of Ti during the deposition operation. Therefore, the composition is controlled by partially covering the alloy target with several pieces of pure Ti.

As-deposited films are amorphous if the substrate temperature is kept below 473 K and do not exhibit shape memory. The amorphous films crystallize when they are annealed at a temperature above 673 K.

Crystallization takes place by nucleation and growth of the B2 phase grains in the amorphous phase. In the course of crystallization, no precipitation occurs in near-equiatomic (50.0–50.5 at% Ni) alloy films but Ti_3Ni_4 precipitates in Ni-rich (greater than 50.6 at%) alloy films. The obtained structures of the near-equiatomic and Ni-rich films are those expected from the information obtained for melt-solidified bulk materials, although the sputter-deposited amorphous films can yield much finer B2 phase grains compared to those of the melt-solidified bulk materials depending on the heat-treatment temperature and time.

Figure 3.39(a)[115] shows an example of an electron diffraction pattern from

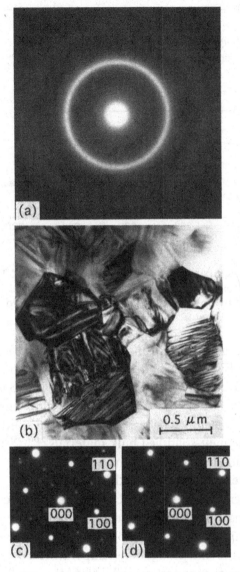

Fig. 3.39. (a) Electron diffraction pattern from an as-sputter-deposited film of a TiNi alloy. (b) Bright field micrograph of a TiNi film annealed at 973 K after sputter deposition (in R-state). (c) Diffraction pattern of (b). (d) Diffraction pattern of the same area but at a slightly higher temperature (in B2 state). (After Gyobu et al.[115])

an as-deposited film of a TiNi alloy. The diffuse rings indicate that the film is amorphous. By annealing the film at 973 K, it crystallizes as shown in Fig. 3.39(b) (a bright field electron micrograph). Figure 3.39(c) is the corresponding diffraction pattern which indicates that the film is in the R-phase state, i.e., the temperature is below R_s. In each grain, variant plates of the R-phase are seen.

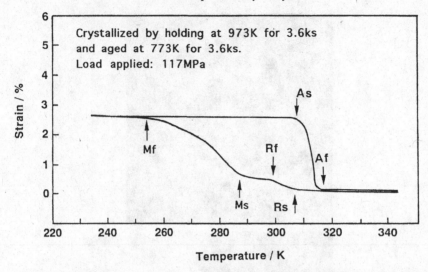

Fig. 3.40. Elongation–temperature curve of a sputter-deposited Ti–51.6% Ni alloy film. Obtained by thermal cycle testing under a constant load of 117 MPa. The film was crystallized by holding at 973 K for 3.6 ks. (After Gyobu et al.[108])

On heating above R_s, the R-phase transforms to the B2 parent as shown in the diffraction pattern in Fig. 3.39(d) and the R-phase plates in Fig. 3.39(b) disappear. The grain size of the B2 phase in Fig. 3.39(b) is 2–3 μm.

As expected from the obtained structures of the films, there is no essential difference in the shape memory behavior between the sputter-deposited crystallized films and the melt-solidified bulk materials.[100–103,108,112] However, reduction of the grain size, which can be achieved in the sputter-deposited crystallized films, is effective for strengthening the matrix and thus for improving the shape memory characteristics. The thermo-mechanical approach (cold-rolling and annealing) which is generally applied for bulk materials of near-equiatomic composition is not applicable in the case of thin films.

Figure 3.40[108] shows the elongation–temperature curves obtained by a constant load thermal cycling test on a Ti–51.6 at% Ni alloy film. It is clear that the film shows a good shape memory effect, which is associated with the two-step transformation, on thermal cycling. The shape memory effect of sputter-deposited Ni-rich films has also been confirmed by conventional tensile tests.[99,106,111]

In contrast to the Ni-rich Ti–Ni alloy films, Ti-rich Ti–Ni alloy films, made by sputter deposition and crystallization, are quite different in structure and resulting properties from the melt-solidified bulk materials.

Ti-rich Ti–Ni alloys crystallized from an amorphous state show a unique microstructure; as a result of an appropriate heat treatment, the amorphous

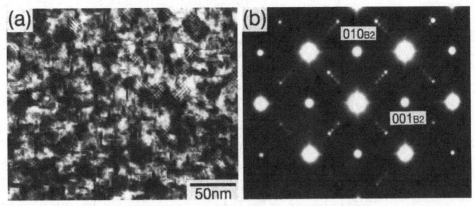

Fig. 3.41. Microstructure of the Ti–48.2 at% Ni alloy film crystallized by holding at 773 K for 3.6 ks. (a) Bright field image. (b) The corresponding diffraction pattern. (After A. Ishida et al.[114])

films yield Guimier–Preston (GP) zones[104,109,110,114] and Ti$_2$Ni precipitates[103,107,108,114] in the process of crystallization of the main mass to the B2 phase. Whether both the precipitates appear or only Ti$_2$Ni appears depends on the heat treatment temperature and the alloy composition. Whether Ti$_2$Ni precipitation occurs prior to, in concurrence with, or after the crystallization of the main mass, also depends on the heat treatment temperature and on the alloy composition.

Ishida et al.[114] demonstrated that a sputter-deposited amorphous Ti–48.2 at% Ni film shows the following precipitation process when crystallized by holding at 773 K: after 300 s of holding, GP zones in a disc shape form; after 3.6 ks, GP zones and spherical particles of Ti$_2$Ni coexist; after 36 ks, GP zones disappear and only Ti$_2$Ni particles remain. Figure 3.41[114] shows the microstructure of the Ti–48.2 at% Ni alloy film crystallized by holding at 773 K for 3.6 ks: (a) is the bright field image of the coexisting GP zones and spherical Ti$_2$Ni precipitates; (b) is the corresponding diffraction pattern. One can see that the GP zones are parallel to the $\{100\}_{B2}$, and that the Ti$_2$Ni precipitates also have a definite orientation relationship with the matrix judging from the alignment of the small Ti$_2$Ni-spots along $[110]_{B2}$ direction.

The GP zones are coherent with the matrix. The existence of the GP zones in the matrix is effective in raising the critical stress for slip deformation and thus improving the shape memory.

Figure 3.42[109] shows strain–temperature curves of a Ti–48.2 at% Ni alloy film crystallized by holding at 745 K for 3.6 ks. The specimen contains GP zones but no Ti$_2$Ni precipitates. The curves were obtained by repeating the thermal cycle test under various constant stresses starting from 30 MPa. The applied stress was increased after each cycle in a stepwise manner up to

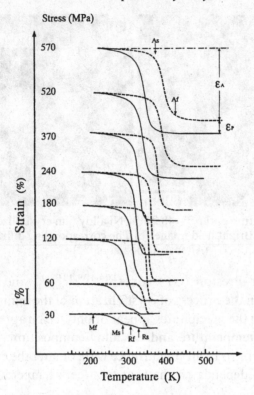

Fig. 3.42. Strain–temperature curves of a Ti–48.2 at% Ni alloy film crystallized by holding at 745 K for 3.6 ks. Obtained by thermal cycle testing under various constant stresses. The specimen contains GP zones but no Ti_2Ni precipitates. (After S. Kajiwara *et al.*[109])

570 MPa. It is seen that recoverable strain increases with increasing applied stress and reaches 5.5% at 240 MPa with no appreciable unrecoverable strain.

The B2–R transformation is important when Ti–Ni films are used for actuation purposes, since temperature hysteresis of this transformation is small and thermal response is good. Composition dependence of the B2 → R transformation temperatures has been examined on a film specimen crystallized from amorphous by holding at 773 K for 3.6 ks.[108] Figure 3.43 shows the obtained result. It is seen that start temperatures of the B2 → R transformation ($\triangle$) and those of the R → B2 transformation ($\blacktriangle$) become maximum (335 K) at the equiatomic composition and decrease with increasing Ni content with a rate of 96 K/at% Ni in the range of 50.0–50.5 at% Ni when heat-treated at 773 K. They are constant (335 K) in the Ti-rich range of the phase diagram of above 50.0 at% Ti. They are also constant in the Ni-rich range of above 50.5 at% Ni (287 K when heat-treated at 773 K).[108]

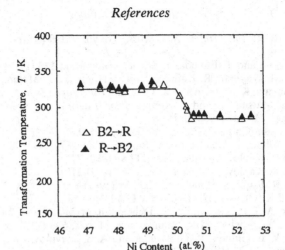

Fig. 3.43. Start temperatures of the B2 → R transformation (△) and those of the R → B2 transformation (▲) of Ti–Ni alloy films crystallized by holding at 773 K for 3.6 ks, plotted against Ni contents. (After A. Gyobu *et al.*[108])

3.8 Melt-spun ribbons of Ti–Ni alloys

Ribbons (typically of 0.03 mm thickness) of the Ti–Ni and Ti–Ni–Cu alloys can be fabricated by the melt-spinning technique. Using this technique, Furuya *et al.*[116,117] successfully fabricated ribbons of the Ti–Ni–Cu alloys of Cu contents exceeding 10 at%, which are too brittle for conventional hot-rolling, and tested the shape memory effect. Transformation hysteresis of the melt-spun Ti–Ni–Cu ribbons decreases with increasing Cu content, in agreement with the bulk materials cut from ingots. The melt-spun ribbons of the Ti–Ni–10 Cu (at%) alloy show much larger recoverable strain than the wires fabricated by the conventional method. This good quality is attributed to the texture and high dislocation density introduced during melt-spinning.[117]

References

1. W. J. Buehler, J. V. Gilfrich and R. C. Wiley, *J. Appl. Phys.*, **34** (1963) 1475.
2. R. J. Wasilewski, *Scr. Metall.*, **5** (1971) 127.
3. P. Duwez and J. L. Taylor, *Trans. AIME*, **188** (1950) 1173.
4. D. M. Poole and W. Hume-Rothery, *J. Inst. Met.*, **83** (1954–55) 473.
5. G. R. Purdy and J. G. Parr, *Trans. Metall. Soc. AIME*, **221** (1961) 636.
6. D. Koskimaki, M. J. Marcinkowski and A. S. Sastri, *Trans. Metall. Soc. AIME*, **245** (1969) 1883.
7. R. J. Wasilewski, S. R. Butler, J. E. Hanlon and D. Worden, *Metall. Trans.*, **2** (1971) 229.
8. S. P. Gupta, K. Makherjee and A. A. Johnson, *Mater. Sci. Eng.*, **11** (1973) 283.
9. G. F. Bastin and G. D. Rieck, *Metall. Trans.*, **5** (1974) 1817.
10. M. Nishida, C. M. Wayman and T. Honma, *Metall. Trans.*, **17A** (1986) 1505.

11. T. Saburi, *Proc. MRS Int. Mtg. on Adv. Mats.*, Tokyo, Vol. 9 (Shape Memory Mater.) (1989) p. 77.
12. T. Saburi, S. Nenno and T. Fukuda, *J. Less-Common Met.*, **125** (1986) 157.
13. M. Nishida, C. M. Wayman, R. Kainuma and T. Honma, *Scr. Metall.*, **20** (1986) 899.
14. T. Tadaki, Y. Nakata, K. Shimizu and K. Otsuka, *Trans. Japan Inst. Metals*, **27** (1986) 731.
15. R. Kainuma, M. Matsumoto and T. Honma, *Proc. Intern. Conf. on Martensitic Transf.*, Nara, (Japan Inst. Metals) (1986) p. 717.
16. T. Saburi, T. Tatsumi and S. Nenno, *J. de Phys.*, **43** (1982) C4-261.
17. S. Miyazaki, Y. Ohmi, K. Otsuka and Y. Suzuki, *J. de Phys.*, **43** (1982) C4-255.
18. H. C. Ling and R. Kaplow, *Metall. Trans.*, **11A** (1980) 77.
19. H. C. Ling and R. Kaplow, *Metall. Trans.*, **12A** (1981) 2101.
20. H. C. Ling and R. Kaplow, *Mater. Sci. Eng.*, **51** (1981) 193.
21. S. Miyazaki and K. Otsuka, *Metall. Trans.*, **17A** (1986) 53.
22. G. D. Sandrock, A. J. Perkins and R. F. Hehemann, *Metall. Trans.*, **2** (1971) 2769.
23. C. M. Wayman, I. Cornelis and K. Shimizu, *Scr. Metall.*, **6** (1972) 115.
24. V. N. Khachin, V. E. Gjunter, V. P. Sivokha and A. S. Savvinov, *Proc. Intern. Conf. on Martensitic Transf.*, MIT, (1979) p. 474.
25. M. Matsumoto and T. Honma, *New Aspects of Martensitic Transformations* (Proc. First Japan Inst. Metals Intern. Symp. on Martensite, Kobe, Japan) (1976) p. 199.
26. C. M. Hwang, M. Meichle, M. B. Salamon and C. M. Wayman, *Philos. Mag. A*, **47** (1983) 9, 31.
27. M. B. Salamon, M. E. Meichle and C. M. Wayman, *Phys. Rev. B*, **31** (1985) 7306.
28. S. M. Shapiro, Y. Noda, Y. Fujii and Y. Yamada, *Phys. Rev. B*, **30** (1984) 4314.
29. S. K. Satija, S. M. Shapiro, M. B. Salamon and C. M. Wayman, *Phys. Rev. B*, **29** (1984) 6031.
30. S. Miyazaki and K. Otsuka, *Philos. Mag. A*, **50** (1984) 393.
31. C. M. Hwang and C. M. Wayman, *Scr. Metall.*, **17** (1983) 381.
32. T. H. Nam, T. Saburi, Y. Kawamura and K. Shimizu, *Mater. Trans. JIM*, **31** (1990) 262.
33. G. Airoldi, S. Besseghini, G. Riva and T. Saburi, *Mater. Trans. JIM*, **35** (1994) 103.
34. T. Fukuda, T. Saburi, K. Doi and S. Nenno, *Mater. Trans. JIM*, **33** (1992) 271.
35. T. Saburi, unpublished work.
36. E. Goo and R. Sinclair, *Acta Metall.*, **33** (1985) 1717.
37. K. Otsuka, T. Sawamura and K. Shimizu, *Phys. Stat. Sol. A*, **5** (1971) 457.
38. R. F. Hehemann and G. D. Sandrock, *Scr. Metall.*, **5** (1971) 801.
39. G. M. Michal and R. Sinclair, *Acta Crystallogr.*, **B37** (1981) 1803.
40. W. Bührer, R. Gotthardt, A. Kulik, O. Mercier and F. Staub, *J. Phys. F: Met. Phys.*, **13** (1983) L77.
41. Y. Kudoh, M. Tokonami, S. Miyazaki and K. Otsuka, *Acta Metall.*, **33** (1985) 2049.
42. T. Saburi and S. Nenno, *Proc. Intern. Conf. on Solid–Solid Phase Transformations*, Pittsburgh, (1981) p. 1455.
43. T. Saburi, M. Yoshida and S. Nenno, *Scr. Metall.*, **18** (1984) 363.
44. S. Miyazaki, S. Kimura, K. Otsuka and Y. Suzuki, *Scr. Metall.*, **18** (1984) 883.
45. T. Hara, T. Ohba, E. Okunishi and K. Otsuka, *Mater. Trans. JIM*, **38** (1997) 11.
46. A. G. Rozner and R. J. Wasilewski, *J. Inst. Metals*, **94** (1966) 169.
47. S. Miyazaki, K. Otsuka and Y. Suzuki, *Scr. Metall.*, **15** (1981) 287.
48. T. Saburi, S. Nenno, Y. Nishimoto and M. Zeniya, *J. Iron and Steel Inst. Japan* (Tetsu-to-Hagane), **72** (1986) 571.
49. L. A. Monasevich, Yu. N. Paskal, V. E. Prib, G. D. Timonin and D. B. Chernov, *Metalloved. Term. Obrab. Met.* **9** (1979) 62.
50. T. Todoroki, H. Tamura and Y. Suzuki, *Proc. Intern. Conf. on Martensitic Trans.*, Nara, (ICOMAT-86) p. 748.
51. T. Todoroki and H. Tamura, *J. Japan Inst. Metals*, **50** (1986) 538.
52. T. Todoroki and H. Tamura, *J. Japan Inst. Metals*, **50** (1986) 546.
53. T. Todoroki, PhD thesis, Tohoku University, Sendai (1986) p. 199.
54. G. B. Stachowiak and P. G. McCormick, *Acta Metall.*, **36** (1988) 291.

55. S. Miyazaki, Y. Igo and K. Otsuka, *Acta Metall.*, **34** (1986) 2045.
56. S. Miyazaki, T. Imai, Y. Igo and K. Otsuka, *Metall. Trans. A*, **17A** (1986) 115.
57. Y. Shugo, S. Hanada and T. Honma, *Bulletin of Research Inst. Mineral and Dressing and Metallurgy* (Tohoku Univ.), **41** (1985) 23.
58. Y. Shugo, K. Yamauchi, R. Miyagawa and T. Honma, *Bulletin of Research Inst. Mineral Dressing and Metallurgy* (Tohoku Univ.), **38** (1982) 11.
59. T. Honma, Y. Shugo and M. Matsumoto, *Bulletin of Research Inst. Mineral Dressing and Metallurgy* (Tohoku Univ.), **28** (1972) 209.
60. K. H. Eckelmeyer, *Scr. Metall.*, **10** (1976) 667.
61. N. M. Matveeva, V. N. Khachin and V. P. Shivokha, in: *Stable and Metastable Phase Equilibrium in Metallic Systems*, ed. M. E. Drits, (Nauka, Moscow, 1985) p. 25 (in Russian).
62. K. N. Melton and O. Mercier, *Metall. Trans.*, **9A** (1978) 1487.
63. Y. Shugo, F. Hasegawa and T. Honma, *Bulletin of Research Inst. Mineral Dressing and Metallurgy* (Tohoku Univ.), **37** (1981) 79.
64. T. Saburi, K. Komatsu, S. Nenno and Y. Watanabe, *J. Less-Common Met.*, **118** (1986) 217.
65. T. Saburi, T. Takagaki, S. Nenno and K. Koshino, *Proc. MRS Int'l Mtg. on Adv. Mats.*, Vol. 9 (Shape Memory Materials), (1988) p. 147.
66. T. H. Nam, T. Saburi and K. Shimizu, *Mater. Trans. JIM*, **31** (1990) 959.
67. T. H. Nam, T. Saburi, Y. Nakata and K. Shimizu, *Mater. Trans. JIM*, **31** (1990) 1050.
68. T. H. Nam, T. Saburi and K. Shimizu, *Mater. Trans. JIM*, **33** (1991) 814.
69. S. Miyazaki, I. Shiota, K. Otsuka and H. Tamura, *Proc. MRS Int'l Mtg. on Adv. Mats.*, Tokyo, Vol. 9 (Shape Memory Materials) (1988) p. 153.
70. W. J. Moberly, T. W. Duerig, J. L. Proft and R. Sinclair, *MRS Symp. Proc.*, Vol. 246, (1992) p. 55.
71. Y. C. Lo, S. K. Wu and H. E. Horng, *Acta Metall. Mater.*, **41** (1993) 747.
72. F. J. J. Vanloo, G. F. Bastin and A. J. H. Leenen, *J. Less-Common Met.*, **57** (1978) 111.
73. K. N. Melton, J. Simpson and T. W. Durerig, *Proc. Int. Conf. on Martensitic Transf.* (ICOMAT-86), Nara, Japan, (1986) p. 105.
74. K. N. Melton, J. L. Proft and T. W. Duerig, *Proc. MRS Int. Mtg. on Adv. Mats.*, Tokyo, Vol. 9 (Shape Memory Materials) (1989) p. 165.
75. M. Piao, S. Miyazaki and K. Otsuka, *Mater. Trans. JIM*, **33** (1992) 337.
76. L. C. Zhao, T. W. Duerig, and C. M. Wayman, *Proc. MRS Int. Mtg. on Adv. Mats.*, Tokyo, Vol. 9 (Shape Memory Materials) (1989) p. 171.
77. L. C. Zhao, T. W. Duerig, S. Justi, K. N. Melton, J. L. Prof, W. Yu and C. M. Wayman, *Scr. Metall.*, **24** (1990) 221.
78. C. S. Zhang, L. C. Zhao, T. W. Duerig and C. M. Wayman, *Scr. Metall.*, **24** (1990), 1807.
79. M. Piao, S. Miyazaki and K. Otsuka, *Mater. Trans. JIM*, **33** (1992) 346.
80. M. Piao, K. Otsuka, S. Miyazaki and H. Horikawa, *Mater. Trans. JIM*, **34** (1993) 919.
81. E. Raub and E. Röschel, *Z. Metallkd.*, **59** (1968) 112.
82. P. Krautwasser, S. Bhan and K. Schubert, *Z. Metallkd.*, **59** (1968) 724.
83. Y. Shugo and T. Honma, *Bulletin of Research Inst. Mineral Dressing and Metallurgy* (Tohoku Univ.), **43** (1987) 128.
84. H. C. Donkersloot and J. H. N. van Vucht, *J. Less-Common Met.*, **20** (1970) 83.
85. N. M. Matveeva, Yu. K. Kovneristyi, A. S. Savinov, V. P. Sivokha and V. N. Khachin, *J. de Phys.*, **43** (1982) C4-249.
86. K. Otsuka, K. Oda, Y. Ueno, M. Piao, T. Ueki and H. Horikawa, *Scr. Metall. et Mater.*, **29** (1993) 1355.
87. D. Golberg, Ya Xu, Y. Murakami, S. Morito, K. Otsuka, T. Ueki and H. Horikawa, *Scr. Metall. et Mater.*, **30** (1994).
88. K. Enami, T. Yoshida and S. Nenno, *Proc. Intern. Conf. on Martensitic Trans.*, Nara (ICOMAT-86) (1986) p. 103.
89. K. Enami, Y. Miyasaka and H. Takakura, *Proc. MRS Int. Mtg. on Adv. Mats.*, Tokyo, Vol. 9 (Shape Memory Materials) (1988) p. 135.
90. Y. Watanabe, T. Saburi, Y. Nakagawa and S. Nenno, *J. Japan Inst. Metals* (Nippon-kingokugakkai-shi), **54** (1990) 861.

 91. S. Miyazaki, K. Otsuka and C. M. Wayman, *Acta Metall.*, **37** (1989) 1873.
 92. M. Nishida, K. Yamauchi, A. Chiba and Y. Higashi, *Proc. Intern. Conf. on Martensitic Trans.*, Monterey (ICOMAT-92) (1993), p. 881.
 93. M. Nishida and T. Honma, *Scr. Metall.*, **18** (1984) 1293.
 94. T. Hoshiya, S. Den, H. Ito, H. Itami and S. Takamura, *Proc. Intern. Conf. on Martensitic Trans.*, Nara (ICOMAT-86) (1986) p. 685.
 95. T. Hoshiya, S. Den, H. Ito, S. Takamura and Y. Ichihashi, *J. Japan Inst. Metals*, **55** (1991) 1054.
 96. T. Hoshiya, F. Takada and Y. Ichihashi, *Mater. Sci. Engineer.*, **A130** (1990) 185.
 97. A. Kimura, H. Tsuruga, T. Morimura, S. Miyazaki and T. Misawa, *Mater. Trans. JIM*, **34** (1993) 1076.
 98. T. Kakuta, K. Ara, H. Tamura and Y. Suzuki, *Proc. MRS Int. Mtg. on Adv. Mats.*, Tokyo, Vol. 9 (Shape Memory Materials) (1989) p. 219.
 99. J. D. Busch, A. D. Johnson, C. H. Lee and D. A. Stevenson, *J. Appl. Phys.*, **68** (1990) 6224.
100. S. Miyazaki, A. Ishida and A. Takei, *Proc. ICOMAT-92* (1993) p. 893.
101. S. Miyazaki and A. Ishida, *Mater. Trans. JIM*, **35** (1994) 14.
102. S. Miyazaki, K. Nomura and A. Ishida, *J. de Physique Colloque IV*, C8, **5** (1995) 677. (Proc. ICOMAT-95).
103. Y. Kawamura, S. Gyobu, H. Horikawa and T. Saburi, *J. de Physique Colloque IV*, C8, **5** (1995) 683. (Proc. ICOMAT-95).
104. Y. Nakata, T. Tadaki, H. Sakamoto, A. Tanaka and K. Shimizu, *J. de Physique Collogue IV*, C8, **5** (1995) 671.
105. M. Kohl, K. D. Shrobanek, E. Quandt, P. Schlossmacher, A. Schussler and D. M. Allen, *J. de Physique Colloque IV*, C8, **5** (1995) 1187. (Proc. ICOMAT-95).
106. Li Hou and D. S. Grummon, *Scr. Metall. et Mater.*, **33** (1995) 989.
107. A. Ishida, M. Sato, A. Takei and S. Miyazaki, *Mater. Trans. JIM*, **36** (1995) 1349.
108. A. Gyobu, Y. Kawamura, H. Horikawa and T. Saburi, *Mater. Trans. JIM*, **37** (1996) 697.
109. S. Kajiwara, T. Kikuchi, K. Ogawa, T. Matsunaga, *Philos. Mag. Letters*, **74** (1996) 137.
110. S. Kajiwara, K. Ogawa, T. Kikuchi, T. Matsunaga and S. Miyazaki, *Philos. Mag. Letters*, **74** (1996) 395.
111. A. Ishida, A. Takei, M. Sato, S. Miyazaki, *Thin Solid Films* **281–282** (1996) 337.
112. A. Ishida, M. Sato, A. Takei, K. Nomura and S. Miyazaki, *Metall. Trans. Mater. A.*, **27A** (1996) 3753.
113. D. S. Grummon and T. J. Pence, *Proc. MRS 1996 Fall Meeting.* (Advances in Materials for Smart System – Fundamentals and Applications)
114. A. Ishida, K. Ogawa, M. Sato and S. Miyazaki, *Metall. Trans.*, **28A** (1997) 1985.
115. A. Gyobu, Y. Kawamura, H. Horikawa and T. Saburi, unpublished work.
116. Y. Furuya, M. Matsumoto, H. S. Kimura, K. Aoki and T. Masumto, *Mater. Trans. JIM*, **31** (1990) 504.
117. Y. Furuya, M. Matsumoto, H. S. Kimura and T. Masumoto, *Mater. Sci. Eng.*, **A147** (1991) L7.

4

Cu-based shape memory alloys

T. TADAKI

Symbols

A elastic anisotropy
A2 body-centered cubic structure
A_f reverse transformation finish temperature
A_s reverse transformation start temperature
B2 CsCl type ordered structure
c_{ij} elastic stiffness
DO$_3$ Cu$_3$Al type ordered structure
k fraction of atoms occupying a specific atomic plane
L1$_0$ CuAu(I) type ordered structure
L2$_1$ Cu$_2$MnAl type ordered structure
M_f martensite finish temperature
M_s martensite start temperature
T_0 equilibrium temperature between parent and martensite

α primary solid solution of Cu, face-centered cubic structure
β high temperature disordered phase, body-centered cubic structure
β_1 DO$_3$ type ordered parent phase
β_2 B2 type ordered parent phase
γ Cu$_5$Zn$_8$ cubic phase
γ_1' 2H (or 2O) type orthorhombic martensite phase
γ_2 Cu$_9$Al$_4$ cubic phase
ζ AgZn ordered hexagonal phase
θ Al$_2$Cu tetragonal phase
σ_c critical stress for slip
2H (or 2O) orthorhombic martensite structure
6R (or 2M) monoclinic martensite structure
18R (or 6M) monoclinic martensite structure

4.1 Phase diagrams of typical Cu-based shape memory alloys

Ti–Ni based shape memory alloys (SMAs) exhibit good properties in strength, ductility and resistance to corrosion, which are important for practical use, in addition to excellent SME characteristics. However, they are very expensive, compared with Cu-based SMAs. Hence, inexpensive Cu-based SMAs have been investigated and developed, which have advantages in electrical and thermal conductivities and deformability, compared with Ti–Ni based SMAs. Cu–Zn based SMAs have actually been used, and Cu–Al based SMAs have shown promise.

The martensitic transformation start temperature, M_s, of the β phase Cu–Zn binary alloys around 40 at% Zn is far below room temperature,[1] and in the alloys with Zn contents lower than about 38 at% a massive transformation occurs, which is a composition invariant but diffusional phase change.[2] Then, ternary elements such as Al, Ga, Si and Sn are added to raise M_s and stabilize the β phase. Among them, Cu–Zn–Al ternary SMAs are the most excellent

97

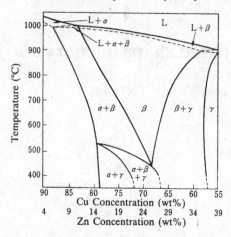

Fig. 4.1. Phase diagram for Cu–Zn–Al ternary system. Vertical cross-section with 6 mass% Al. (After L. Delaey *et al.*[3])

with respect to ductility and grain boundary fracture. A vertical section of the equilibrium phase diagram of the Cu–Zn–Al ternary alloy system at a constant Al content of 6 mass% is shown in Fig. 4.1[3] (the ordinate is graduated in Celsius according to the original figure). Here, the high temperature β phase is disordered bcc, but upon quenching to ambient temperature ordering occurs into B2 or D0$_3$ (or L2$_1$) type structure, and then transforms into 9R (or 6M according to a new notation by Otsuka *et al.*[4]) or 18R(6M) martensite with or without further cooling, depending on alloy composition. When the Al content becomes higher, the β phase is susceptible to decomposition into α (the fcc primary solid solution of Cu) and γ (Cu$_5$Zn$_8$ cubic phase) at around 700 K.

The β phase in the binary Cu–Al alloy system undergoes a eutectoid decomposition at 838 K into α and γ_2 (Cu$_9$Al$_4$ cubic phase), but if it is quenched from the β phase region to ambient temperature, it transforms martensitically.[5] The martensite phases are disordered 9R (or 6M) for less than 11 mass% Al, 18R (or 6M) for 11–13 mass% Al and 2H (or 2O) for more than 13 mass% Al.[6] The β phase with more than 11 mass% Al becomes ordered before the martensitic transformation takes place. The M_s temperature is still above room temperature even at around 14 mass% Al, and the alloys with higher Al contents are susceptible to precipitation of γ_2 and do not undergo the martensitic transformation. When a ternary element Ni is added, the precipitation of γ_2 is effectively suppressed because diffusion of Cu and Al is inhibited by Ni. Excellent SMEs are observed in alloys with Al content close to 14 mass% and with a varying Ni content. However, since the alloys become brittle with increasing Ni content,[7] the optimum compositions lie around Cu–14 ~ 14.5 Al–3 ~ 4.5 Ni (mass%). Figure 4.2[8] shows a vertical section of the equilib-

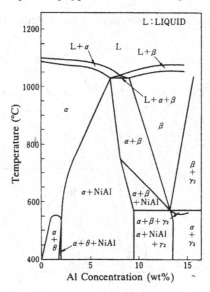

Fig. 4.2. Phase diagram for Cu–Al–Ni ternary system. Vertical cross-section with 3 mass% Ni. (After D. P. Dunne and N. F. Kennon[8])

rium phase diagram of the Cu–Al–Ni ternary alloy system at 3 mass% Ni (the ordinate is graduated in Celsius according to the original figure). In the figure, NiAl is an equilibrium phase of B2 type as a product of the decomposition of β. The martensites formed in the ternary SMAs are basically the same as those in the binary alloys, depending on Al content.

The M_s temperature of the ternary Cu–Zn–Al alloys is shown in Fig. 4.3[9] as a function of composition (here, temperature is graduated in Celsius according to the original figure). It is thus possible to adjust the M_s temperature of the SMAs roughly between 173 ∼ 473 K by changing composition. The martensitic transformation temperatures of Cu–xAl–4.0 Ni (mass%) SMAs are shown in Fig. 4.4[10] as a function of x (the ordinate is graduated in Celsius according to the original figure). It is seen that the transformation temperatures decrease steeply with slight increase of x. Although they also decrease with increasing Ni content when the Al content is fixed, the dependence of M_s upon Ni is much weaker than that of Al, and the decrease with increasing Ni content is effectively due to the increase of Al content.

It should be noted here that although the main factor controlling the transformation temperatures is alloy composition, they are also significantly affected by other factors such as heat-treatment, quenching rate, grain size and the number of transformation cycles. For example, in general the smaller the grain size the lower the M_s temperature.[11] Effects of the other factors will be explained later.

Martensitic Transformation Start Temperature (°C)

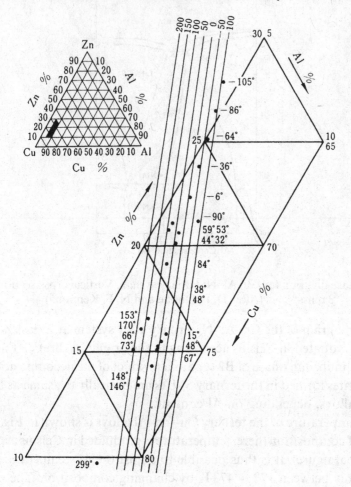

Fig. 4.3. Relation between M_s temperature and alloy composition in Cu–Zn–Al SMAs. (After L. Delaey et al.[9])

4.2 Mechanical behavior

4.2.1 Shape memory effect and superelasticity in polycrystalline state

Single-crystalline Cu-based SMAs exhibit unusual mechanical behavior, which is far beyond common sense for metals and alloys, as shown in Chapter 2. Mechanical behavior of the polycrystalline Cu-based SMAs, however, is different from that of the single-crystalline ones. Figure 4.5[12] shows an example of stress–strain (S–S) curves of a polycrystalline Cu–Al–Ni SMA under tensile stress. One of the characteristic properties of the polycrystalline SMAs is that the recoverable strains due to SME are considerably reduced, as

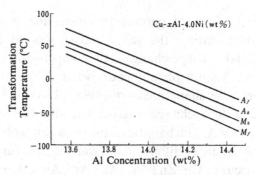

Fig. 4.4. Relation between M_s temperature and Al content in Cu–Al–Ni SMAs. (Courtesy of Prof. K. Otsuka[10])

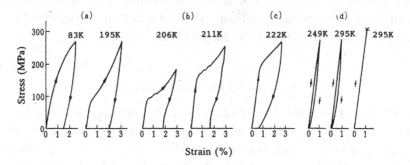

Fig. 4.5. Deformation behavior of polycrystalline Cu–Al–Ni SMA; $M_s = 191$ K, $M_f = 203$ K, $A_s = 213$ K, $A_f = 225$ K. (After K. Oishi and L. C. Brown[12])

seen in (a)–(c). Generally speaking, this is because the transformation strains generated are different from grain to grain and then averaged over the whole crystal, and grain boundary fracture easily occurs. In this case the strain recoverable due to superelasticity (SE) at temperatures above A_f is substantially zero, as seen in (d). The reason for this is as follows: although the fracture stress in the single crystals is about 600 MPa,[13] that in the polycrystals is only about 280 MPa, as seen in (d); grain boundary fracture then occurs before the martensites are stress-induced. The ease of grain boundary fracture in Cu–Al–Ni SMAs is thus a serious problem to be solved for practical use.

Cu–Zn–Al SMAs are ductile even in the polycrystalline state, and thus exhibit SE even after a stress of about 300 MPa is applied.[14] This is because of the ordered structure of the parent phase, as will be explained later.

4.2.2 Reverse shape memory effect

When a SMA is heavily deformed at a temperture below or slightly above M_s and then heated to a temperature above A_f, the shape recovery due to SME becomes incomplete because of permanent strain due to slip introduced. However, with further heating, the shape moves toward the direction of applied stress. This kind of shape change is time-dependent, while that due to SME occurs instantly. Such a phenomenon was first found in a Cu–Zn–Si SMA and termed 'reverse shape memory effect' (RSME).[15] It was demonstrated that the reverse shape change is associated with the bainitic transformation occurring in the SMA. The bainitic transformation is observed to exhibit a dual nature, characteristic of diffusion and diffusionless (martensitic) transformations, and to occur in Cu–Zn (–Si, –Al, –Au), Ag–Cd and Ag–Zn SMAs when heat-treated at around 473 K.[16] Although the mechanism of the transformation is still in dispute, it is believed that the shape change associated with the transformation interacts with internal or external stress, and that this interaction results in macroscopic shape change.[17]

4.2.3 Grain boundary fracture

Polycrystalline Cu-based SMAs are so brittle that they are susceptible to intergranular cracking (IGC) even upon quenching from betatizing temperature, although they exhibit outstanding SE as large as 18% in the single-crystalline state.[18] Many potential applications of these SMAs are thus limited by this intergranular embrittlement. IGC is not due to the segregation of impurities at grain boundaries such as Bi, Sb, S, P, O and Pb, which are known to cause intergranular weakness in Cu.[19] It is likely that the abnormally high elastic anisotropy ($A = 2c_{44}/(c_{11} - c_{12}) = 13$ for Cu–Al–Ni SMAs[20] and 15 for Cu–Zn–Al SMAs,[21] where c_{ij} are elastic stiffness) leads to stress concentration at grain boundaries due to elastic and plastic strain incompatibilities between grains.[22] The typically large grain sizes (≈ 1 mm) contribute to the stress concentration as well. The critical yield stress is in general higher for the ordered alloys than for disordered ones. For example, it is ~ 600 MPa for a Cu–Al–Ni SMA with D0$_3$ (or L2$_1$) order[23] and ~ 200 MPa for a Cu–Zn–Al SMA with B2 order in the parent phase.[24] B2 ordered alloys are comparatively easy to plastically deform, but anomalous strengthening occurs during the B2 → D0$_3$ and B2 → L2$_1$ ordering in Fe$_3$Al[25] and Ag$_2$MgZn[26] alloys, respectively.

However, a detailed examination[27] has demonstrated that the most crucial cause of IGC in polycrystalline Cu–Al–Ni SMAs is the formation of stress-induced martensites along grain boundaries upon quenching. If there is no deformation mode accommodating the stress due to the displacement associated with the formation of a stress-induced martensite to maintain compatibility at a grain boundary, the displacement causes a crack of a width depending on the difference of orientations between two adjacent grains. Upon quenching the cracks easily form when the difference between the temperature of the quenching medium and the M_s temperature is large, because otherwise stress-induced martensites would form uniformly in all grains. Then, in order to prevent IGC effectively, it is necessary to reduce the size of the cracks produced by the displacements, and/or to accommodate the stress produced by the formation of stress-induced martensites. Favorable conditions for the reduction of the crack size are small grain size, small orientation dependence of transformation strain and ease of plastic deformation. Taking account of these factors, problems such as why In–Tl SMAs exhibit good resistance against IGC despite their extremely large elastic anisotropy, can be reasonably elucidated.[27]

4.2.4 Fatigue fracture

When SMAs as well as other alloys are subjected to cyclic deformation, dislocations are usually introduced into them, and they gradually fatigue and eventually fracture. Since fatigue restricts the use of SMAs as functional devices, its origin must be clarified to promote their utilization.

When a single-crystalline Cu–Al–Ni SMA is repeatedly deformed in the SIM mode above the A_f temperature, the shape of the superelastic loop in an S–S curve is stable and almost unchanged up to about 10^3 cycles. This is because the alloy with DO_3 (or $L2_1$) order has a high critical stress for slip, σ_c, and thus dislocations are difficult to introduce during cyclic deformation.[28]

In general, however, the shape of the S–S curve changes with cyclic deformation because of the introduction of slip.[29] In polycrystals, in particular, slip relaxes stress concentration at grain boundaries due to elastic and/or plastic strain incompatibility between neighboring grains. Even at temperatures above the A_f temperature, stress-induced martensites are stabilized by the stress field of the dislocations, resulting in residual strains upon unloading. On subsequent loading, the formation of stress-induced martensites is assisted by the stress field, and thus the critical stress for the formation of stress-induced martensites decreases.

Fatigue life of Cu-based SMAs strongly depends on the deformation mode. Figure 4.6[30] shows relations between stress and number of cycles (S–N relation) of Cu–Al–Ni SMAs, of which single crystals and polycrystals are subjected to cyclic deformation with three kinds of deformation mode, i.e., elastic deformation in the parent state (a), deformation due to SIM (b) and deformation due to rearrangement of 2H (2O) martensite (c). In Cu–Al–Ni SMAs with high σ_c, the fatigue life in the polycrystalline state is shorter than in the

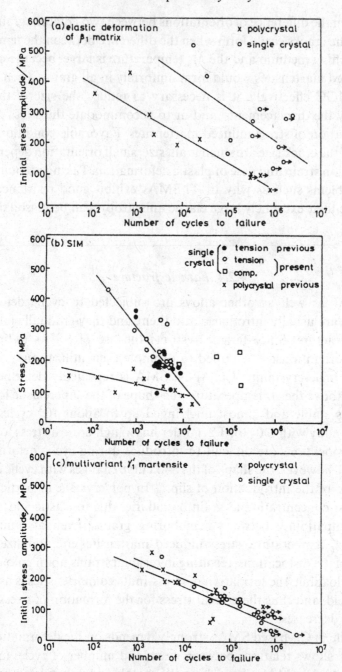

Fig. 4.6. *S–N* relation for single- and polycrystalline Cu–Al–Ni SMAs; (a) when elastically deformed in parent phase, (b) when deformed by the formation of stress-induced martensites in parent phase and (c) when deformed in multivariant 2H(2O) martensites. (After H. Sakamoto[30])

single-crystalline state, irrespective of the deformation mode. In particular, the fatigue life of the polycrystalline Cu–Al–Ni SMAs is about two to four orders shorter than those of the single-crystalline ones in elastic deformation and SIM modes, respectively. This is because slip stress is higher and thus IGC takes place more easily in the polycrystalline state. On the other hand, the difference between single- and polycrystalline states is small for the martensite reorientation mode because many twinning systems are operable in the martensitic state,[31] and hence the strain incompatibilities between neighboring grains can be accommodated by twinning. The fatigue life of the Cu–Al–Ni SMAs thus becomes shorter in the order of elastic deformation, reorientation of martensite variants, and SIM, i.e., the deformation mode associated with the movement of parent/martensite interfaces has the worst effect on the fatigue life.[30]

4.3 Aging effects of shape memory alloys

4.3.1 In the parent phase

The martensitic transformations from metastable parent to martensitic phases are essential for the occurrence of SME and SE. Hence it is rather natural that the SME characteristics, such as the transformation temperatures, the thermal hysteresis, the recoverable strain, the transformation stress during the shape recovery, etc., are more or less subject to aging effects in both phases. In fact, various aging phenomena occur in Cu-based,[32–35] Ti–Ni based,[36] Ni–Al based[37] and Fe–Ni based SMAs.[38] Among them Cu-based SMAs are, in general, prone to aging effects even at ambient temperature, and this is an important issue to be solved in the field of materials science.

Upon aging Cu-based SMAs in the parent phase at elevated temperatures, the precipitation of the equilibrium α, γ (Cu_5Zn_8), γ_2 (Cu_9Al_4) or NiAl (B2 type) phase is expected to take place from their phase diagrams, as shown in Figs. 4.1 and 4.2. The precipitation depresses the martensitic transformation initially due to coherency strain fields around precipitates and/or due to the destruction of potential nucleation sites of martensites. However, accompanying the growth of the precipitates, solute atoms in the parent phase are enriched or depleted, depending on the chemical compositions of the precipitates. As a result, the M_s temperature eventually decreases or increases, and concurrently the shape recovery becomes worse in both cases. Preceding the formation of the equilibrium phase, the bainitic transformation occurs most typically in some Cu–Zn based SMAs, and it is associated with the RSME, as mentioned before.

Aging also affects the ordering from the high temperature β phases (A2) to

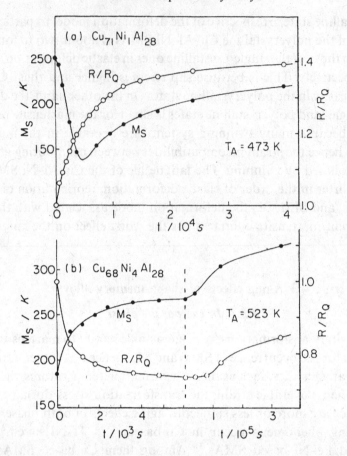

Fig. 4.7. Changes in electrical resistivity and M_s temperature upon aging Cu–Al–Ni SMAs at (a) 473 K and (b) 523 K. (After F. Nakamura et al.[39])

B2, DO_3 or $L2_1$ type structures. Although the A2 → B2 ordering cannot be suppressed by quenching because of its second order nature, the subsequent B2 → DO_3 (or $L2_1$) ordering is often incomplete during quenching in some SMAs because of its first order nature. Besides, excess quenched-in vacancies are introduced in the parent phases during quenching. Then, upon aging at intermediate temperatures, the ordering reaction from incomplete to complete order becomes dominant, being assisted by the migration of quenched-in vacancies. The martensitic transformation temperatures are strongly affected by the ordering reactions, i.e., raised or depressed, depending on alloy system and chemical composition. In Fig. 4.7[39] electrical resistivity versus temperature curves are shown, which reveal how M_s temperature is affected by precipitation and ordering upon aging the parent phase of Cu–Al–Ni SMAs. The SMAs with lower Ni content are susceptible to decomposition and thus the precipitation of the γ_2 phase proceeds upon aging, as seen from the increase of electrical resistivity in (a). Accompanying the increase of electrical resistivity, the M_s temperature varies in a complicated manner with aging, as in some Fe–Ni based ternary alloys.[40] On the other hand, in SMAs with higher Ni content, the electrical resistivity decreases in the early stage of aging, as shown in (b), indicating that the ordering reaction proceeds. Thus the increase of M_s temperature is caused by the ordering reaction. In fact, the increase in M_s temperature during aging is attributed to the progress of ordering between next nearest neighbors (nnn) in the L2$_1$ parent lattice[41] by means of AL-CHEMI (Atom Location by Channeling Enhanced Microanalysis),[42] which has been insufficient in the as-quenched state. It is also identified by AL-CHEMI that one of the two Cu sites in the D0$_3$ parent lattice, i.e., the body-center sites referring to the fundamental bcc lattice, are preferentially occupied by Ni atoms, irrespective of heat treatment.[41] In this sense, the parent lattice is strictly of L2$_1$ type. The later stage of aging in Fig. 4.7 is associated with precipitation.

A similar correlation between M_s and nnn ordering is found in Cu–Zn–Al SMAs.[43] However, the M_s temperature in another Cu–Zn–Al[44] and also a Au–Cu–Zn SMA[45] is observed to be almost unchanged during aging. These experimental facts suggest that the kinetics of ordering in the parent phase depends on not only the alloy system but also its chemical composition.

Whether the ordered structure at lower temperature in Cu–Zn–Al SMAs was D0$_3$ or L2$_1$ had not been conclusive, but it has recently been solved by neutron diffraction.[46] The parent phase possesses an L2$_1$ type ordered structure with the following distribution of constituent atoms into the sublattices: one half of the corner sites referring to its fundamental bcc lattice are occupied by Al and Zn atoms, the remaining by Cu and Zn atoms and all of the body-center sites by Cu atoms.

4.3.2 In the martensite phase

Aging in the martensitic state also brings about drastic changes in the SME characteristics. The aging effects are termed 'stabilization of martensite', because the A_s temperature of some SMAs is increased with aging time and eventually the reverse transformation is suppressed by aging. The higher the aging temperature the faster the increase in A_s temperature, as seen from Fig. 4.8,[44] which is for a Cu–Zn–Al SMA. Moreover, when quenching is carried out stepwise in the parent phase, the stabilization is markedly retarded. These facts indicate that the stabilization of martensite is a vacancy-assisted phenomenon.[47]

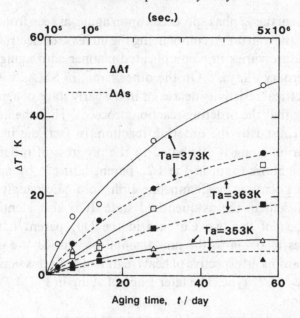

Fig. 4.8. Changes of A_s and A_f, i.e., ΔA_s and ΔA_f, respectively, in Cu–Zn–Al SMA as a function of aging time, where T_a is aging temperature. (After Y. Nakata et al.[44])

Primarily two mechanisms have been proposed for the stabilization of martensite. One is reordering in martensite, i.e., atomic rearrangement in martensite resulting in some change in the relative stability between parent and martensite.[48,49] The reordering in martensite is suggested by the fact that the angular separations between some reflections such as $12\bar{2}$ and 202 from 18R (6M) martensites, which reflect the orthorhombic distortion in the basal plane of the martensite due to atomic ordering, certainly decrease with aging time.[50] Thus the reordering in martensite is regarded as a kind of disordering.

Another mechanism for the stabilization of martensite is pinning of interfaces between parent and martensite and between martensite variants by quenched-in vacancies and/or precipitates.[49,51] However, the stabilization of martensite occurs even in single-crystalline states of martensite.[52] The pinning mechanism is thus not always valid.

A notable phenomenon concurring with the stabilization of martensite is the rubberlike behavior of the martensitic SMAs. The mechanism is not clear at present, as mentioned in Chapter 2. In recent works[44,53–55] the rubberlike behavior was expected to be closely related to some changes in crystal structure which cause the stabilization of martensite. Very recently, however, by means of precise X-ray diffraction techniques, no significant changes in integrated intensities and peak positions were found for various kinds of martensites in some SMAs,[56–59] including the trigonal martensites in near equiatomic

Au–Cd SMAs, which were recently found to exhibit rubberlike behavior upon martensite aging.[60] Then, some mechanisms taking account of this experimental fact were proposed anew for the rubberlike behavior, as discussed in Chapter 2.

Whatever the cause, the stabilization of martensites can be substantially suppressed by hot-rolling for a Cu–Zn–Al alloy. It was demonstrated[61] that once hot-rolled, the martensitic Cu–Zn–Al SMA still exhibits the reverse transformation into the parent phase at about 363 K upon heating, even after aging at room temperature for 1 year. This is to be noted because the same SMA step-quenched and then aged at room temperature for 2 weeks or water-quenched after betatization does not undergo the reverse transformation at all. Although the origin is not understood yet, it is suggested that the formation of dislocations with a high density at high temperature during rolling is a key factor in providing high mobility pathways and sinks for vacancies. It is to be noted that if the dislocations are those introduced by low temperature fatigue, the stabilization behavior of martensite is not improved. It thus appears essential for the dislocations to be introduced by hot-rolling.

4.4 Thermal cycling effects

Thermal cycling, by which one forward and reverse transformation cycle is completed, eventually produces a large number of dislocations in parent phases of SMAs. Figure 4.9[62] shows transmission electron micrographs indicating that the density of dislocations increases with the number of thermal cycles in a Cu–Al–Ni SMA. The accumulated dislocations cause shifts of the transformation temperatures, as shown in Fig. 4.10.[62] The Cu–Al–Ni SMA was subjected to thermal cycling by which the DO_3 (or $L2_1$) $\leftrightarrow$ 2H(2O) transformations were repeated. The M_s temperature decreases markedly with increasing number of thermal cycle, as does the A_f temperature. Since the changes of both M_s and A_f temperatures are parallel, they are attributed to that of the equilibrium temperature T_0 between parent and martensite. Meanwhile, it is found by X-ray diffraction that the intensities of superlattice reflections corresponding to the nnn order in the DO_3 (or $L2_1$) parent phase significantly decrease due to thermal cycling, while those corresponding to the nn order are hardly affected thereby.[62] Therefore the decrease in T_0 temperature brought about by thermal cycling is due to the decrease in the nnn order in the $L2_1$ parent phase.

The above fact suggests that the thermal cycling effects may be different between SMAs with B2 and DO_3 (or $L2_1$) order. In fact, the density of dislocations is much higher in SMAs with B2 order than those with DO_3 (or

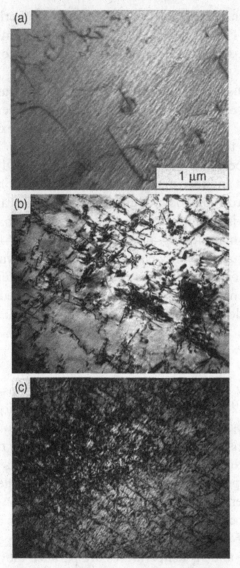

Fig. 4.9. Transmission electron micrographs showing dislocated substructure of the parent phase of Cu–Al–Ni SMA after (a) 1, (b) 10^3 and (c) 10^4 thermal cycles. (After Y. Nakata et al.[62])

$L2_1$) order.[63] In addition, the M_s temperature is depressed by thermal cycling in Cu–Zn–Al SMAs with $D0_3$ (or $L2_1$) order, as in the Cu–Al–Ni SMA, while it is raised in those with B2 order. However, an X-ray diffraction work[63] has shown that in the Cu–Zn–Al SMA with $D0_3$ (or $L2_1$) order not only the nnn type superlattice reflections but also the nn type ones decrease in intensity in a similar manner with increasing number of thermal cycle.

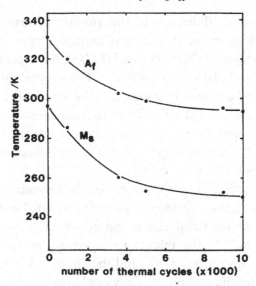

Fig. 4.10. Variation in M_s and A_f temperatures as a function of the number of thermal cycles in Cu–Al–Ni SMA. (After Y. Nakata *et al.*[62])

The Burgers vectors of the dislocations left in the parent phase are thus expected to be $1/4\langle 111 \rangle$ and $1/2\langle 111 \rangle$ in the Cu–Zn–Al SMAs with DO_3 (or $L2_1$) order and B2 order, respectively. The $1/4\langle 111 \rangle$ dislocations destroy not only the nnn type order but also the nn order in the Cu–Zn–Al SMAs with DO_3 (or $L2_1$) order, and the $1/2\langle 111 \rangle$ dislocations destroy the nn order in the Cu–Zn–Al SMAs with B2 order. However, these Burgers vectors inferred from the X-ray diffraction work are not consistent with those deduced from transmission electron microscopy (TEM) analysis, i.e., $\langle 100 \rangle$ for Cu–Zn(–Al) SMAs with B2 order, [64,65] which is equivalent to $1/2\langle 100 \rangle$ in the DO_3 (or $L2_1$) type lattice. Although the results inferred from the X-ray diffraction work cannot preclude the local existence of the $1/2\langle 100 \rangle$ or $\langle 100 \rangle$ type dislocations observed by TEM, the actually observed disordering characteristics cannot be understood unless $1/4\langle 111 \rangle$ or $1/2\langle 111 \rangle$ type dislocations are present with fairly large amounts in the parent phases of SMAs with DO_3 (or $L2_1$) or B2 order, respectively.

The density of dislocations accumulated by thermal cycling also depends on the transformation mode.[66] The amount of dislocations produced by the repetition of the DO_3 (or $L2_1$) $\leftrightarrow$ 18R (6M) transformations is larger than that by the repetition of the DO_3 (or $L2_1$) $\leftrightarrow$ 2H (2O) transformations. This fact indicates that the relaxation of stress due to the volume change is not the primary cause for the production of dislocations, because the volume change associated with the two transformation modes should be the same to the first approximation. Then the difference in the amount of dislocations may be attributed to those in atomistic structures of martensite/parent interfaces and intervariant ones between 18R (6M) and 2H (2O) martensites.

Local stresses around dislocations may assist or suppress the nucleation of martensite. On the other hand, martensites once formed may be assisted to revert to the parent phase or stabilized by the local stresses. Anyhow, after a large number of thermal cycles, the transformation temperature intervals, $M_s - M_f$ and $A_f - A_s$, increase in general due to the stress fields produced by the abundance of dislocations.

Several mechanisms have been proposed for the formation of dislocations with specific Burgers vectors, e.g. the reaction of partial dislocations associated with stacking faults on the basal planes and non-basal planes of 9R (6M) or 18R (6M) martensites,[64,67] the relaxation of stress due to volume change accompanied by the transformation[65] and the coalescence between different plate-groups of martensite variants.[68] They are likely to occur, but it appears that the observed thermal cycle characteristics cannot be explained consistently by any one of them.

4.5   Improvements of shape memory alloys

4.5.1   Grain-refined SMAs

Various attempts have been made to improve the thermal and mechanical stability of the SME characteristics, especially for the most promising Cu–Zn–Al and Cu–Al–Ni SMAs. Pre-aging or step-quenching is effective for retardation of aging effects because the excess quenched-in vacancies are thereby annealed out to some extent. However, one of the most noteworthy improvements to be made for Cu-based SMAs is in their fatigue and fracture characteristics because they are fatal defects of the SMAs. As mentioned above, the ease of fracture and fatigue is due to the difficulty of relaxation of stress concentrations at grain boundaries. Hence it becomes essentially important to refine the grain size of these SMAs for their improvement.

Grain refining by small additions of alloying elements is being attempted most intensively. The effects of various elements such as Ti, Zr, V, Pb, B etc. have been examined in Cu–Al–Ni and Cu–Zn–Al SMAs. Among them, the influence of Ti additions to Cu–Al–Ni SMAs is to be noted in particular, because the grain size is reduced down to 15 μm, while the usual size is roughly 1 mm.[69] The mechanism of grain refinement was formerly accounted for by the presence of finely dispersed Ti-rich 'X-phase' particles that act as obstacles to grain boundary migration;[70] or it was suggested[69,71] that although Ti atoms are concentrated primarily in the X-phase, the grain growth retardation is due to the small concentrations of Ti atoms in the solid solution. However, the mechanism appeared to be not so simple. Two kinds of precipitates were found to be present in the Ti-doped SMA, termed X_L and X_S phases.[72] The X_L and X_S precipitates, typically several μm and several tens nm in size, respectively, were both identified as possessing $L2_1$ type $(Cu,Ni)_2TiAl$ structures, but with different proportions of the constituent atoms. Then it was proposed[73] that grain refinement is the combined result of two effects: the presence of Ti atoms brings about lowered diffusion rates of constituent atoms, resulting in grain refinement upon casting; and the pinning effects of the X_S particles suppress the growth of β grains, resulting in the ultimate refinement.

By grain refining, the fracture strength, the strain to fracture and the fatigue life were all improved to a great extent. For example, when the grain size is refined to less than $20\,\mu$m, the number of cycles to failure for Ti-doped Cu–Al–Ni SMA increases as much as that for Ti–Ni based SMAs, i.e., 10^5, as shown in Fig. 4.11.[74]

Other methods for grain refining have been tried, such as thermomechanical treatment,[75,76] mechanical alloying,[77] rapid solidification,[78] sputtering[79] and powder metallurgy.[80,81] For example, Cu–Al–Ni SMAs with grain sizes as

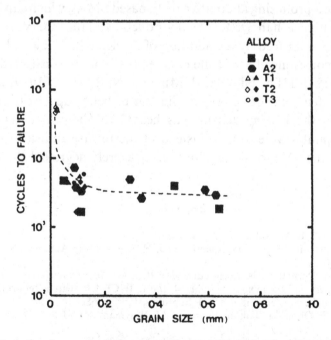

Fig. 4.11. Variation in number of stress cycles to failure as a function of grain size in Cu–Al–Ni SMAs; alloys A1 and A2 contain no Ti and alloys T1, T2 and T3 contain Ti. (After G. N. Sure and L. C. Brown[74])

small as 5 μm were obtained by thermomechanical treatment,[76] and as a result, a fracture stress of 1200 MPa and a fracture strain as high as 10% were obtained.

4.5.2 High temperature shape memory alloys

Cu–Al–Ni SMAs exhibit much better thermal stability than other Cu-based ones. However, they had not been put to practical use because of their lack of ductility. Then, Cu–Al–Ni–Ti–Mn SMAs were developed as high temperature SMAs, which could be used above 373 K.[82] The Ti addition has the effect of grain refinement, as mentioned above, and the Mn addition strongly influences the transformation temperatures as Al does so. Meanwhile, the ductility of Cu–Al–Ni SMAs significantly decreases with a small increment of Al content. Therefore, the replacement of Al with Mn effectively improves the ductility substantially with no change in transformation temperature. It was shown that a Cu–Al–Ni–Mn–Ti SMA with the eutectoid composition exhibits the best thermal resistance for the SME: no precipitation occurs in it upon aging at 623 K for up to roughly 5 h.[83]

Although Ni–Al based SMAs are not included in Cu-based alloys, they are expected to be a promising alternative to Ti-based SMAs, which can be used at temperatures higher than 373 K. The poor ductility of the binary Ni–Al SMAs was drastically improved by the addition of 3d elements.[84] Two-phase Ni–Al based SMAs consisting of β (B2) and ductile γ (fcc) phases were thus developed, such as Ni–25Al–15Fe, Ni–11Al–30Mn and Ni–19Al–12Mn–9Fe (at%).[85] However, the temperature of use was limited to below approximately 523 K because when the L1$_0$ martensite was heated to above that temperature, Ni$_5$Al$_3$ precipitates were formed instead of the reverse transformation. The ductility and the SME of these alloys were severely degraded thereby.[86]

References

1. H. Pops and T. B. Massalski, *Trans. AIME*, **230** (1964) 1662.
2. T. B. Massalski, in *Phase Transformations*, ed. H. I. Aaronson (American Society for Metals, 1970) p. 433.
3. L. Delaey, A. Deruyttere, N. Aernoudt and J. R. Roos, *Shape Memory Effect, Super-Elasticity and Damping in Cu–Zn–Al Alloys*, INCRA Research Report (Project No. 238) (Katholieke Universiteit Leuven), February, 1978) p. 68.
4. K. Otsuka, T. Ohba, M. Tokonami and C. M. Wayman, *Scr. Metall. Mater.*, **29** (1993) 1359.
5. P. R. Swann and H. Warlimont, *Acta Metall.*, **11** (1963) 511.
6. Z. Nishiyama, *Martensitic Transformation* (Academic Press, New York, 1970) p. 80.
7. S. W. Husain and P. C. Clapp, *J. Mater. Sci.*, **22** (1987) 509.
8. D. P. Dunne and N. F. Kennon, *Metals Forum*, **4** (1981) 176.

9. L. Delaey, A. Deruyttere, N. Aernoudt and J. R. Roos, *Shape Memory Effect, Super-Elasticity and Damping in Cu–Zn–Al Alloys,* INCRA Research Report (Project No. 238) (Katholieke Universiteit Leuven), February, 1978) p. 76.
10. S. Miyazaki, S. Ichinose and K. Otsuka, unpublished work, University of Tsukuba, Japan (1983).
11. Z. Nishiyama, *Martensitic Transformation* (Academic Press, New York, 1970) p. 283.
12. K. Oishi and L. C. Brown, *Metall. Trans.,* **2** (1971) 1971.
13. S. Miyazaki, T. Kawai and K. Otsuka, *Proc. Int. Conf. Martensitic Transformations (ICOMAT-82),* Leuven, Belgium, *J. de Phys.,* **43** (Suppl. 4) (1982) C4-813.
14. L. Delaey, A. Deruyttere, N. Aernoudt and J. R. Roos, *Shape Memory Effect, Super-Elasticity and Damping in Cu–Zn–Al Alloys,* INCRA Research Report (Project No. 238) (Katholieke Universiteit Leuven), February, 1978) p. 91.
15. H. Pops, in *Shape Memory Effects,* ed. J. Perkins (Plenum Press, New York, 1975) p. 525.
16. For example, Y. Nakata, T. Tadaki and K. Shimizu, *Mater. Trans. JIM,* **30** (1989) 107.
17. K. Takezawa and S. Sato, *Trans. JIM,* **11** (1988) 894.
18. K. Otsuka, H. Sakamoto and K. Shimizu, *Acta Metall.,* **27** (1979) 585.
19. S. W. Husain and P. C. Clapp, *J. Mater. Sci.,* **22** (1987) 2351.
20. M. Suezawa and K. Sumino, *Scr. Metall.,* **10** (1976) 789.
21. G. Guénin, M. Morin, P. F. Gobin, W. Dejonghe and L. Delaey, *Scr. Metall.,* **11** (1977) 1071.
22. S. Miyazaki, K. Otsuka, H. Sakamoto and K. Shimizu, *Trans. Jpn. Inst. Metals,* **22** (1981) 244.
23. S. Miyazaki, T. Kawai and K. Otsuka, *Scr. Metall.,* **16** (1982) 431.
24. K. Takezawa, H. Waizumi and S. Sato, *Bulletin of the Faculty of Engineering,* Hokkaido Univ., No. 125 (1985) 191.
25. N. S. Stoloff and R. G. Davis, *Acta Metall.,* **12** (1964) 473.
26. M. Yamaguchi and Y. Umakoshi, *Scr. Metall.,* **15** (1981) 605.
27. H. Sakamoto and K. Shimizu, *Trans. Jpn. Inst. Metals,* **27** (1986) 601.
28. H. Sakamoto, K. Shimizu and K. Otsuka, *Trans. Jpn. Inst. Met.,* **22** (1981) 579.
29. L. C. Brown, *Metall. Trans. A.,* **13A** (1982) 25.
30. H. Sakamoto, *Trans. Jpn. Inst. Met.,* **24** (1983) 665.
31. K. Otsuka, *Proc. Int. Conf. Martensitic Transformations (ICOMAT-89),* Sydney, Australia, *Mater. Sci. Forum,* **56–58** (1990) 393.
32. M. Ahlers, *Proc. Int. Conf. Martensitic Transformations (ICOMAT-86),* Nara, Japan, *Jpn. Inst. Metals* (1986) p. 786.
33. T. Honma, S. Miyazaki and K. Otsuka, in *Type and Mechanical Characteristics of Shape Memory Alloys,* ed. H. Funakubo (Gordon and Breach Science Publishers, New York, 1987) p. 61.
34. T. Tadaki, K. Otsuka and K. Shimizu, *Ann. Rev. Mater. Sci.,* **18** (1988) 25.
35. T. Tadaki and K. Shimizu, *MRS Int'l. Mtg. on Adv. Mats.,* **9** (1989) 291.
36. M. Nishida and T. Honma, in Ref. [13], p. C4-225.
37. K. Enami, in Ref. [13], p. C4-727.
38. T. Maki, K. Kobayashi, M. Minato and I. Tamura, *Scr. Metall.,* **18** (1984) 1105.
39. F. Nakamura, J. Kusui, Y. Shimizu and J. Takamura, *J. Jpn. Inst. Metals,* **44** (1980) 1302 (in Japanese).
40. E. Hornbogen and W. Meyer, *Acta Metall.,* **15** (1967) 584.
41. Y. Nakata, T. Tadaki and K. Shimizu, *Mater. Trans. JIM,* **31** (1990) 652.
42. J. C. H. Spence and J. Taftø, *J. Microsc.,* **130** (1983) 147.
43. T. Suzuki, Y. Fujii and A. Nagasawa, in Ref. [31], p. 481.
44. Y. Nakata, O. Yamamoto and K. Shimizu, *Mater. Trans. JIM,* **34** (1993) 429.
45. T. Tadaki, H. Okazaki, Y. Nakata and K. Shimizu, *Mater. Trans. JIM,* **31** (1990) 935.
46. A. Planes, L. Mañosa, E. Vies, J. Rodríguez-Carvajal, M. Morin, G. Guénin and J. L. Macqueron, *J. Phys.: Condens. Matter.,* **4** (1992) 553.
47. J. van Humbeeck, J. Janssen, Mwamba-Ngoie and L. Delaey, *Scr. Metall.,* **18** (1984) 893.
48. J. Janssen, J. van Humbeeck, M. Chandrasekaran, N. Mwamba and L. Delaey, in Ref. [13], C4-715.

49. G. Scarsbrook, J. Cook and W. M. Stobbs, in Ref. [13], p. C4-703.
50. L. Delaey, T. Suzuki and J. van Humbeeck, *Scr. Metall.*, **18** (1984) 899.
51. A. Abu Arab and M. Ahlers, in Ref. [13], p. C4-709.
52. A. Abu Arab, M. Chandrasekaran and M. Ahlers, *Scr. Metall.*, **18** (1984) 709.
53. A. Abu Arab and M. Ahlers, *Acta Metall.*, **36** (1988) 2627.
54. M. H. Wu and C. M. Wayman, *Materials Science Forum*, **56–58** (1990) 553.
55. T. Tadaki, H. Okazaki, Y. Nakata and K. Shimizu, *Mater. Trans. JIM*, **31** (1990) 941.
56. J. Ye, Doctoral Thesis, Univ. of Tokyo (1990) p. 77.
57. T. Ohba, K. Otsuka and S. Sasaki, in Ref. [31], p. 317.
58. T. Ohba, T. Finlayson and K. Otsuka, *Suppl. J. de Phys. III*, **5** (1995) C8.1083.
59. Y. Murakami, Y. Nakajima, K. Otsuka and T. Ohba, *Suppl. J. de Phys. III*, **5** (1995) C8.1071.
60. Y. Nakajima, S. Aoki, K. Otsuka and T. Ohba, *Mater. Lett.*, **21** (1994) 271.
61. X. Duan and W. M. Stobbs, *Scr. Metall.*, **23** (1989) 441.
62. Y. Nakata, T. Tadaki and K. Shimizu, *Trans. Jpn. Inst. Metals*, **26** (1985) 646.
63. T. Tadaki, M. Takamori and K. Shimizu, *Trans. Jpn. Inst. Metals*, **28** (1987) 120.
64. D. Ríos-Jara and G. Guénin, *Acta Metall.*, **35** (1987) 121.
65. K. Marukawa and S. Kajiwara, *Philos. Mag. A.*, **A55** (1987) 85.
66. H. Sakamoto, K. Sugimoto, Y. Nakamura, A. Tanaka and K. Shimizu, *Mater. Trans. JIM*, **32** (1991) 128.
67. K. Adachi and J. Perkins, *Metall. Trans. A*, **16A** (1985) 1551.
68. K. Adachi, J. Perkins and C. M. Wayman, *Acta Metall.*, **36** (1988) 1343.
69. G. N. Sure and L. C. Brown, *Metall. Trans. A*, **15A** (1984) 1613.
70. K. Sugimoto, K. Kamei, H. Matsumoto, S. Komatsu, K. Akamatsu, T. Sugimoto, in Ref. [13], p. C4-609.
71. J. S. Lee and C. M. Wayman, *Trans. Jpn. Inst. Met.*, **27** (1986) 584.
72. K. Adachi, Y. Hamada and Y. Tagawa, *Scr. Metall.*, **21** (1987) 453.
73. K. Adachi, K. Shoji and Y. Hamada, *ISIJ International*, **29** (1989) 378.
74. G. N. Sure and L. C. Brown, *Scr. Metall.*, **19** (1985) 401.
75. R. Elst, J. van Humbeeck and L. Delaey, *Mater. Sci. Tech.*, **4** (1988) 644.
76. K. Mukunthan and L. C. Brown, *Metall. Trans. A.*, **19A** (1988) 2921.
77. T. Kaneyoshi, T. Takahashi, Y. Hayashi and M. Motoyama, *J. Jpn. Inst. Metals*, **56** (1992) 517.
78. J. H. Zhu, D. P. Dunne, G. W. Delamore and N. F. Kennon, *Proc. Int. Conf. on Martensitic Transformations* (ICOMAT-92), Monterey Institute for Advanced Studies, Monterey, California, USA, (1993) p. 1083.
79. T. Minemura, H. Andoh, Y. Kita and I. Ikuta, *J. Mater. Sci. Lett.*, **4** (1985) 793.
80. J. Janssen, F. Willems, B. Verelst, J. Maertens and L. Delaey, in Ref. [13], p. C4-809.
81. N. Nakanishi and T. Shigematsu, *Bull. Jpn. Inst. Metals*, **28** (1989) 659 (in Japanese).
82. K. Sugimoto, K. Kamei, H. Matumoto, S. Komatsu and T. Sugimoto, in Ref. [13], p. C4-761.
83. Y. Itsumi, Y. Miyamoto, T. Takashima, K. Kamei and K. Sugimoto, in Ref. [31], p. 469.
84. K. Ishida, R. Kainuma, N. Ueno and T. Nishizawa, *Metall. Trans. A*, **22A** (1991) 441.
85. R. Kainuma, H. Nakano, K. Oikawa, K. Ishida and T. Nishizawa, *Mat. Res. Soc. Symp. Proc.*, **246** (1992) 403.
86. J. H. Yang and C. M. Wayman, *Mater. Sci. Eng.*, **A160** (1993) 241.

5

Ferrous shape memory alloys

T. MAKI

5.1 Morphology and substructure of ferrous martensite

In ferrous alloys, an FCC austenite (γ) is transformed to three kinds of martensites with different crystal structures depending on alloying elements and compositions, i.e., (1) $\gamma \to \alpha'$ (BCC or BCT) martensite, (2) $\gamma \to \varepsilon$ (HCP) martensite and (3) $\gamma \to$ FCT martensite. The most popular ferrous martensite is α' formed in Fe–C and Fe–Ni alloys, etc. The ε martensite is formed only in ferrous alloys with a low stacking fault energy of austenite such as Fe–Cr–Ni and Fe-high Mn alloys. FCT martensite is very rare and has been found only in Fe–Pd and Fe–Pt alloys. A shape memory effect has been observed in each of these three types of ferrous martensites.

The morphology of ε and FCT martensites is a parallel-sided thin plate type with planar interfaces. On the other hand, in the case of α' martensite, five types of morphologies have been reported so far. Figure 5.1[1] shows various morphologies of α' martensite and ε martensite. Five types of α' martensites are distinguished not only morphologically but also crystallographically. Among these α' martensites, lath martensite forms at the highest temperatures and the thin plate martensite forms at the lowest temperatures. With a decrease in M_s temperature, the substructure of martensite changes from dislocated (lath martensite) to twinned (thin plate martensite). The factors that determine the morphology and substructure of α' martensite remain poorly defined. However, the M_s temperature, the relative strengths of austenite and martensite, the critical resolved shear stress for slip and twinning in martensite, and the stacking fault energy of austenite are considered to be the important factors.[2-6]

Lath and lenticular are two major morphologies of α' martensite.[4] Lath martensite is formed in Fe–C ($< 0.6\%$ C) and Fe–Ni ($< 28\%$ Ni) alloys and most heat-treatable commercial steels, and has overwhelming industrial significance because it is a basic structure of high strength steels. Lenticular

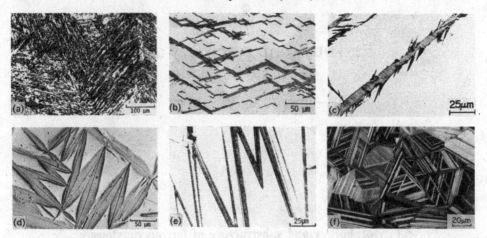

Fig. 5.1. Optical micrographs of various types of martensites in ferrous alloys: (a) lath
α′ (Fe–7%Ni–0.22%C), (b) butterfly α′ (Fe–20%Ni–0.73%C), (c) (225)$_A$ type plate α′
(Fe–8%Cr–0.92%C), (d) lenticular α′ (Fe–29%Ni–0.26%C), (e) thin plate α′
(Fe–31%Ni–0.23%C), (f) ε martensite (Fe–24%Mn). (After Maki[1])

martensite appears in Fe–C (0.8 ~ 1.8%) and Fe–Ni (29 ~ 33%) alloys. The
other three α′ martensites are not common in ferrous alloys. However, among
the five types of α′ martensites, only the thin plate martensite has the possibility
of exhibiting a perfect shape memory effect, as will be shown later.

5.2 Ferrous alloys exhibiting shape memory effect

Table 5.1 summarizes the ferrous alloys exhibiting a perfect or nearly perfect
shape memory effect which have been reported so far in the literature. All
ferrous shape memory alloys except Fe–Pt alloy have disordered structure
(FCC) in the parent phase. Fe–Pt,[21] Fe–Pd[22] and Fe–Ni–Co–Ti[10] alloys
undergo a thermoelastic martensitic transformation with a small thermal
hysteresis when the austenite is properly heat treated. The other alloys such as
Fe–Ni–C and Fe–Mn–Si–(Cr) exhibit a non-thermoelastic martensite.

In the case of α′ (BCC or BCT) martensite, Wayman[7] first found that an
Fe–25 at% Pt alloy shows a perfect shape memory effect when the austenite
matrix is ordered (L1$_2$). This was the first ferrous shape memory alloy. In this
alloy, the martensitic transformation changes from a non-thermoelastic (burst
type) to a thermoelastic type with an increase in the degree of long range order
of austenite.[21] Furthermore, Koval et al.[9] have reported that an
Fe–23% Ni–10% Co–10% Ti alloy shows a shape memory effect when the
austenite is aged. Maki et al.[10] have also found that the properly ausaged

Fe–33% Ni–10% Co–4% Ti alloy containing fine coherent ordered precipitates (γ'-Ni$_3$Ti (L1$_2$ type)) exhibits a thermoelastic martensite and hence a perfect shape memory effect and superelasticity. Jost[23] has found that the ausaged Fe–32% Ni–12% Co–4% Ti alloy shows a two-way shape memory effect. Kajiwara[12] has reported that a nearly perfect shape memory effect appears in an Fe–31% Ni–0.4% C alloy when the austenite is strengthened by ausforming although the martensite of this alloy is not a thermoelastic type. Recently, it has been reported by Koval and Monastyrsky[13] that the ausaged Fe–31% Ni–7% Nb alloy which contains γ''-Ni$_3$Nb (D0$_{22}$ type) particles also exhibits a perfect shape memory effect.

The Fe–Mn–Si shape memory alloy, which undergoes $\gamma \to \varepsilon$ martensitic transformation, was first discovered by Sato et al.[14] They found that a single crystal of an Fe–30% Mn–1% Si alloy exhibits a complete shape memory effect when it is deformed in a specific direction, and clearly showed that the origin of the shape memory effect is a reverse transformation of stress-induced ε. Sato et al.[24] and Murakami et al.[15] have also succeeded in developing the polycrystalline Fe–Mn–Si shape memory alloys. Moriya et al.[16] and Otsuka et al.[17] have developed the stainless type Fe–Cr–Ni–Mn–Si–(Co) shape memory alloys which have good corrosion resistance. Recently, Tsuzaki et al.[18] have reported that the addition of C to Fe–Mn–Si alloys is very effective for improvement of shape memory effect. In the case of ε martensite, the transformation is not a thermoelastic type and hence the thermal hysteresis of forward and reverse transformations is fairly large.

In the case of FCT martensite, Oshima et al.[19,22] have found that the successive martensitic transformation from FCC $\to$ FCT $\to$ BCT occurs in an Fe– $\sim$ 30 at% Pd alloy and shown that the FCC $\to$ FCT transformation is thermoelastic accompanying the shape memory effect. They[20] have also found the shape memory effect associated with the thermoelastic FCC $\to$ FCT martensitic transformation in the ordered Fe–25 at% Pt alloy.

The origin of shape memory effect associated with non-thermoelastic martensite is the reverse transformation of stress-induced martensites. In this case, there are two necessary conditions in order to obtain a perfect shape memory effect. One is that the deformation takes place only by the stress-induced martensitic transformation without accompanying a slip. Therefore, it is desirable that the yield stress of the austenite matrix is as high as possible. The second condition is that the shape strain of stress-induced martensite is completely reversible. For this, it is necessary that the martensite interface remains mobile and the forward and reverse transformations occur by the movement of the martensite interface. A planar martensite interface seems to be necessary for reversible movement.[25] Actually, the martensite morphology of all the

Table 5.1. Ferrous alloys exhibiting a perfect or nearly perfect shape memory effect

Crystal structure of martensite	Alloy	Composition	Nature of transformation*	M_s (K)	A_s (K)	A_f (K)	Ref.
BCC or BCT (α')	Fe–Pt (ordered γ)	≈ 25 at%Pt	TE	131	—	148	[7],[8]
	Fe–Ni–Co–Ti (ausaged γ)	23%Ni–10%Co–10%Ti	—	173	243	≈ 443	[9]
		33%Ni–10%Co–4%Ti	TE	146	122	219	[10]
		31%Ni–10%Co–3%Ti	Non-TE	193	343	508	[11]
	Fe–Ni–C (ausformed γ)	31%Ni–0.4%C	Non-TE	<77	—	≈ 400	[12]
	Fe–Ni–Nb (ausaged γ)	31%Ni–7%Nb	Non-TE	≈ 160	—	—	[13]
HCP (ε)	Fe–Mn–Si	30%Mn–1%Si (single crystal) (28 ~ 33)%Mn–(4 ~ 6)%Si	Non-TE	≈ 300	≈ 410	—	[14]
	Fe–Cr–Ni–Mn–Si	9%Cr–5%Ni–14%Mn–6%Si	Non-TE	≈ 320	≈ 390	≈ 450	[15]
		13%Cr–6%Ni–8%Mn–6%Si–12%Co	Non-TE	≈ 293	≈ 343	≈ 573	[16]
		8%Cr–5%Ni–20%Mn–5%Si 12%Cr–5%Ni–16%Mn–5%Si	Non-TE	≈ 260	≈ 370	< 573	[17]
	Fe–Mn–Si–C	17%Mn–6%Si–0.3%C	Non-TE	323	453	494	[18]
FCT	Fe–Pd	≈ 30 at%Pd	TE	179	—	183	[19]
	Fe–Pt	≈ 25 at%Pt	TE	—	—	300	[20]

*TE: Thermoelastic martensite, Non-TE: Non-thermoelastic martensite

ferrous shape memory alloys in Table 5.1 is a thin plate type with a planar interface.

5.3 Shape memory effect associated with α' thin plate martensite

5.3.1 Characteristics of thin plate martensite

It is to be noted that, among various α' martensites, only the thin plate martensite exhibits a perfect shape memory effect. This martensite is character-ized by a highly smooth and planar interface, and transformation twins in martensite extend completely from one interface to the other (Fig. 5.2[26]). Furthermore, the austenite in the vicinity of martensite rarely contains disloca-tions, indicating that the transformation strain is elastically accommodated in austenite. This is in contrast to observations made on the other types of α' martensites. The interface of thin plate martensite remains mobile irrespective of the thermal hysteresis and thus a martensite plate gradually thickens during cooling and shrinks during heating by movement of the interface.[25-27] This nature is not observed in other types of α' martensites. The existence of fully extended transformation twins in martensite might be important for the reversible movement of the interface.

Judging from the characteristics of thin plate α' martensite, i.e., completely twinned and perfectly elastic accommodation in austenite, the following four factors might be favorable for the formation of thin plate martensite[27] in ferrous alloys: (1) a high strength of austenite, (2) a small transformation volume change and a small magnitude of transformation shape strain, (3) a low M_s temperature and (4) a high tetragonality of martensite. A high tetragonality

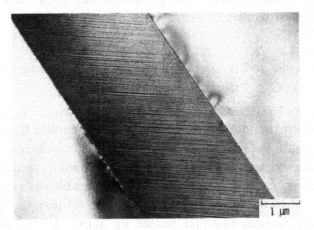

Fig. 5.2. Transmission electron micrograph of thin plate martensite in Fe–30%Ni–0.39%C alloy ($M_s = 133$ K). (After Maki *et al.*[26])

of martensite corresponds to a small twinning shear, a low magnitude of shape strain and a low twin boundary energy.[28,29]

Since the interface of thin plate martensite is mobile, this martensite possesses the necessary conditions for thermoelastic martensite. However, thin plate α' martensite is not always thermoelastic, e.g., the thermal hysteresis (the temperature difference between M_s and A_f) of the thin plate martensite in the Fe–Ni–C alloy shown in Fig. 5.2 is large, at about 600 K. Then, in order for the thin plate α' martensite to become thermoelastic, the introduction of suitable factors for reducing its thermal hysteresis is necessary.

Judging from various previous observations,[25-32] it can be considered that the important condition for reducing the thermal hysteresis of thin plate martensite is to reduce the thickness of transformation twins in martensite. A high strength of austenite and a high tetragonality of BCT martensite seem to be favorable for reducing the twin thickness in martensite. Actually, the ordered Fe–Pt alloy[21,30,33] and the ausaged Fe–Ni–Co–Ti alloy,[10] which exhibit a thermoelastic martensite, satisfy the conditions described above.

5.3.2 Fe–Ni–Co–Ti shape memory alloys

When the Fe–Ni–Co–Ti alloys are ausaged under the proper conditions, fine particles of ordered γ'-Ni_3Ti ($L1_2$ type) are precipitated coherently in the austenite matrix, and the alloys exhibit a thin plate martensite.[10,11,34] The thermal hysteresis of these thin plate martensites changes widely with the chemical composition and ausaging conditions. However, thin plate martensite shows a perfect shape memory effect irrespective of the amount of thermal hysteresis. Two typical examples of shape memory effect associated with α' thin plate martensites with different thermal hysteresis in Fe–Ni–Co–Ti alloys are shown here.

5.3.2.1 Shape memory effect associated with thin plate martensite exhibiting a large thermal hysteresis

When an Fe–31% Ni–10% Co–3% Ti alloy is ausaged at 873 K for 3.6 ks after solution treatment, the alloy exhibits a thin plate martensite (M_s, A_s and A_f are 193 K, 343 K and 508 K, respectively). Figure 5.3[11] shows the relation between the degree of shape recovery and the deformation (bending) temperature in the ausaged specimen. In this test, the ausaged specimen which was fully austenitic at room temperature was bent (the maximum strain of the specimen was 2.5%) at various temperatures, and then heated up above A_f. The perfect shape memory effect is obtained when the specimen is deformed at temperatures around M_S. This shape memory effect is caused by reverse transformation of

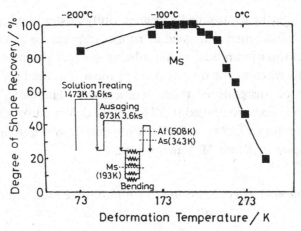

Fig. 5.3. Effect of deformation (bending) temperature on the degree of shape recovery after heating above A_f in Fe–31%Ni–10%Co–3%Ti alloy ausaged at 873 K for 3.6 ks.[11]

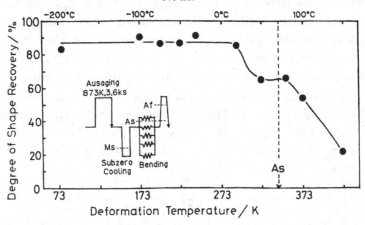

Fig. 5.4. Effect of deformation (bending) temperature on the degree of shape recovery after heating above A_f in Fe–31%Ni–10%Co–3%Ti alloy ausaged at 873 K for 3.6 ks and then subzero cooled at 77 K. (After Maki *et al.*[11])

stress-induced thin plate martensite. A good shape memory effect is also obtained by deformation at 77 K at which the specimen contains a large amount of martensite (about 70%) before deformation. At this temperature, deformation occurs mainly by thickening of the existing martensite plates and the movement of α'/α' interfaces.

When the ausaged specimen is once subzero cooled to 77 K, a nearly perfect shape memory effect is obtained by deformation over a wide temperature range between room temperature and 77 K as shown in Fig. 5.4.[11] It is interesting that the subzero-cooled specimen containing about 70% marten-site exhibits a good shape memory effect even when deformed at room temperature, although the as-ausaged specimen (fully austenitic) hardly shows a shape memory effect when deformed at room temperature as can be seen in Fig. 5.3. The deformation modes of the subzero-cooled specimen are the same as those at 77 K, whereas the deformation of the as-ausaged specimen occurs mainly by slip of austenite at room temperature. Therefore, the ausaged specimen which is subzero cooled to 77 K before deformation shows a nearly perfect shape memory effect after deformation even at room temperature (well above M_s) in spite of a low M_s temperature.

5.3.2.2 Shape memory effect associated with thin plate martensite exhibiting a small thermal hysteresis

When an Fe–33% Ni–10% Co–3.5% Ti–1.5% Al alloy is ausaged at 973 K for 3.6 ks, the specimen exhibits a typical thermoelastic thin plate martensite with a small thermal hysteresis ($M_s = 155$ K, $A_s = 126$ K and $A_f = 204$ K).[1] When this specimen is deformed at temperatures around M_s, thermoelastic martensite with a small thermal hysteresis is stress-induced. Even when the specimen is deformed at temperatures around room temperature (well above M_s), thin plate martensite is induced. However, the stress-induced martensite formed at room temperature is non-thermoelastic with a large thermal hysteresis ($A_s = \sim 700$ K and $A_f = \sim 1000$ K). When the specimen is deformed at temperatures between around M_s (155 K) and about 253 K, both types of martensites, i.e., thermoelastic and non-thermoelastic thin plate martensites are stress-induced.

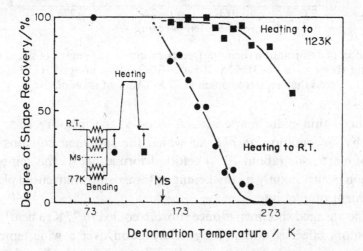

Fig. 5.5. Effect of deformation (bending) temperature on the degree of shape recovery after heating up to room temperature or 1123 K in Fe–33%Ni–10%Co–3.5%Ti–1.5%Al alloy ausaged at 973 K for 3.6 ks. (After Maki[1])

Shape memory effect in this ausaged specimen is summarized in Fig. 5.5[1] which shows the relation between the deformation (bending) temperature and the degree of shape recovery on heating up to room temperature or 1123 K (above A_f of the stress-induced non-thermoelastic martensite). When deformed at 77 K, the specimen exhibits a perfect shape memory effect on heating up to room temperature. On the other hand, when the specimen is deformed at around room temperature, a nearly perfect shape memory effect appears at very high temperatures whilst heating the specimen. Therefore, the ausaged specimen in this alloy can be used as a shape memory alloy which operates both at low temperature (below room temperature) and at very high temperature (about 1000 K). When the specimen is deformed at temperatures between M_s and 257 K, a shape recovery occurs partly during heating to room temperature due to the reverse transformation of stress-induced thermoelastic martensite and a further recovery occurs by further heating up to a temperature above A_f of the stress-induced non-thermoelastic martensite.

5.3.2.3 Role of γ′ precipitates and Co addition on the martensitic transformation in Fe–Ni–Co–Ti alloys

The precipitation of γ'-Ni_3Ti particles causes various changes in austenite hardness, M_s temperature, martensite morphology, etc.[34]

Austenite matrix is age-hardened by the precipitation of γ'. The γ' particles in austenite are inherited by the martensite. When the austenite containing fine coherent FCC particles transforms to martensite, there are two alternatives:[35] (1) the particles transform with the matrix (remain coherent) to a BCC (or BCT) structure, (2) the particles remain FCC and an incoherent interface between martensite and particles is created (Fig. 5.6). Transition from 1 to 2 depends on the particle size. As the particles reach a critical size, case 2 will apply instead of case 1. For case 1, the martensite in Fe–Ni–Ti–(Co) alloy exhibits a tetragonality,[36] because γ' is the ordered structure ($L1_2$). This is the case for the ausaged Fe–Ni–Co–Ti shape memory alloys described in the previous sections. On the other hand, when γ' particles are inherited by martensite without shearing (case 2), the martensite exhibits no tetragonality and the formation of transformation twins is remarkably suppressed.[37]

In the case of Fe–Ni–Ti ternary alloys, lenticular martensite is usually formed even in the ausaged specimen. However, the Fe–Ni–Co–Ti alloys easily produce thin plate martensite by ausaging.[34] By the addition of Co, the transformation volume change is decreased. This is due to the Invar effect.[38] Austenite hardness is also increased by the addition of Co. This might arise from the increase in the amount of γ' precipitates by Co addition.[39]

As described above, all of the factors which are favorable for the formation

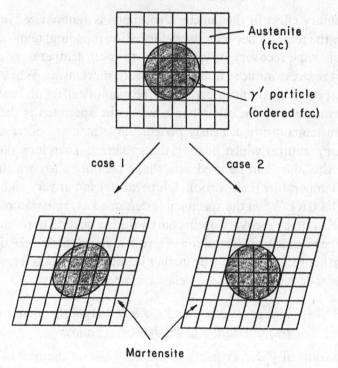

Fig. 5.6. Schematic illustration of the martensitic transformation of austenite containing coherent particles.

of thin plate martensite and for reducing the thermal hysteresis of thin plate martensite described in Section 5.3.1 can be obtained by the precipitation of γ' particles in austenite and the addition of Co to Fe–Ni–Ti alloys.

5.4 Shape memory effect associated with ε martensite in Fe–Mn–Si alloys

5.4.1 Fe–Mn–Si shape memory alloys

Among various ferrous shape memory alloys, the cost-saving Fe–Mn–Si–(Cr)–(Ni) alloys appear to have commercial significance. Murakami et al.[15] showed that Fe–Mn–Si alloys with 28 ~ 33% Mn and 4 ~ 6% Si exhibit a nearly perfect shape memory effect as shown in Fig. 5.7. The addition of Si seems to be essential to obtain a perfect shape memory effect in Fe-high Mn alloys. In these alloys, the specimen is a metastable austenite structure at room temperature and the stress-induced ε martensite is easily formed without accompanying a slip by deformation at room temperature.

Otsuka et al.[40] have found that the shape memory effect of Fe–Mn–Si alloys is increased remarkably by repetition of slight deformation (about 2.5%) and annealing at about 873 K (this treatment was called 'training'). Figure 5.8[40] shows the effect of 'training' on the shape memory effect. After five cycles of training, a perfect shape memory effect is achieved. This treatment, 'training', suppresses slip deformation through introducing dislocations which raise the strength of the austenite matrix, and generates martensite at lower stress through introducing stacking faults in austenite which act as nucleation sites for ε martensite.

5.4.2 General features of γ → ε martensitic transformation and ε → γ reverse transformation

The $\gamma \to \varepsilon$ martensitic transformation proceeds by the motion of one $a/6\langle 112 \rangle$ Shockley partial dislocation on every second (111) austenite plane (Fig. 5.9). In this case, there are three possible shear directions per $(111)_\gamma$ plane. These three kinds of shear system are equivalent and lead to HCP crystals with the same orientation. The $\gamma \to \varepsilon$ transformation with only one kind of shear direction causes a total shear of 0.353. This transformation shear is much larger than that of the $\gamma \to \alpha'$ martensitic transformation (about 0.2).

The thermally-induced or spontaneous transformation of ε martensite is known to be generally self-accommodating.[41-44] The ε band usually consists of three laminated variants with different shear directions in order to minimize the total shape strain of the transformed region. In contrast, the stress-induced ε transformation is generally accomplished by the selective motion of a single

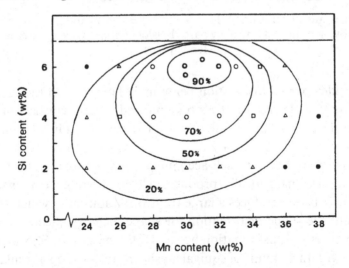

Fig. 5.7. Chemical composition dependence of shape memory effect determined by room temperature bending in Fe–Mn–Si alloys. (After Murakami *et al.*[15])

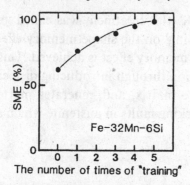

Fig. 5.8. Effect of 'training' on the degree of shape memory effect in
Fe–32%Mn–6%Si alloy. (After Otsuka et al.[40])

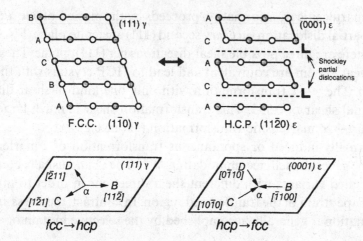

Fig. 5.9. Schematic illustration of the $\gamma \rightarrow \varepsilon$ transformation and $\varepsilon \rightarrow \gamma$ reverse transformation by operation of Shockley partial dislocatgions on every alternate {111}
austenite plane.

type of Shockley partial dislocation being most favorable to the direction of
applied stress.[42,43] Therefore, the stress-induced ε band consists of a single
variant. Figure 5.10 schematically shows a thermally-induced band and a
stress-induced ε band. In the case of the thermally-induced ε band, the total
shape strain is almost zero by self-accommodation of three kinds of variants,
although each laminate locally produces a large homogeneous shear. The
stress-induced ε band produces a large transformation shear which effectively
contributes to the shape deformation for shape memory effect.

The $\varepsilon \rightarrow \gamma$ reverse transformation occurs by the motion of Shockley partial
dislocations with three kinds of equivalent shear directions in a similar way to
the $\gamma \rightarrow \varepsilon$ forward transformation (Fig. 5.9). These three kinds of shear directions per $(0001)_\varepsilon$ plane lead to an FCC crystal with the same orientation.

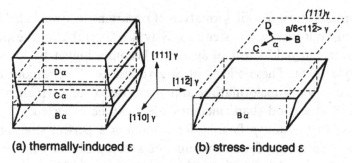

Fig. 5.10. Schematic illustration of the shape change of (a) thermally-induced and (b) stress-induced ε bands.

However, it should be emphasized that the transformation strain by ε marten-site is completely recovered only when the partial dislocations operating on forward transformation move in reverse upon reverse transformation.[42–45] In the case of Fe–high Mn binary alloys, the $\varepsilon \to \gamma$ reverse transformation occurs by the operation of three kinds of equivalent partial dislocations. However, in the case of Fe–Mn–Si shape memory alloys, the reverse transformation occurs by the reversible movement of one kind of transformation dislocations. This reversible movement of transformation dislocations in Fe–Mn–Si alloys has been confirmed by surface relief observation of ε martensite upon heating above the A_f temperature.[44] The surface relief of ε martensite in Fe–Mn–Si shape memory alloys almost completely disappears with the reverse trans-formation, although in the case of Fe–Mn binary alloys most of the surface relief of ε martensite remains even after the reverse transformation.

As to the cyclic transformation behavior in Fe–Mn–Si shape memory alloys, it has been reported that the $\gamma \to \varepsilon$ transformation is enhanced by thermal cycling between room temperature and 573 K.[46,47]

5.4.3 Role of Si in shape memory effect in Fe–high Mn alloys

In order to obtain a perfect shape memory effect in Fe–high Mn alloys, the addition of the proper amount of Si is necessary. The origin of the shape memory effect in Fe–Mn–Si alloys is the reverse transformation of stress-induced ε martensite. Thus, a shape change must take place only by the stress-induced $\gamma \to \varepsilon$ martensitic transformation without accompanying a slip. From this point of view, the effect of Si has been discussed in connection with the solution hardening of austenite, the decrease in the Néel point and the stacking fault energy of austenite.[15] The decrease in Néel point of austenite by the addition of Si increases the driving force for the $\gamma \to \varepsilon$ transformation, and makes the critical stress for the formation of ε martensite lower. The decrease in the stacking fault energy of austenite by Si is also favorable for the formation of ε. Moreover, the yield stress of austenite becomes higher by the solution hardening due to Si. These effects of Si give rise to the promotion of the stress-induced $\gamma \rightarrow \varepsilon$ transformation and the absence of permanent slip in austenite, so that a good shape memory effect can be expected.

In addition to the above factors, for obtaining a perfect shape memory recovery, the reverse transformation must be accomplished by the backward movement of the partial dislocations which operated in the forward transformation. As was described in the previous section, the addition of Si has an effect which makes the movement of partial dislocations reversible during forward and reverse transformation. This important role of Si can be emphasized by the fact that the Fe–24% Mn binary alloy does not exhibit a good shape memory effect, even when a shape change takes place only by the stress-induced ε transformation without accompanying a slip.[48]

Since three $1/6\langle 112\rangle$ type partial dislocations are equivalently possible to move on $\{111\}$ for the $\gamma \leftrightarrow \varepsilon$ transformations, a special condition must be met for the reversible movement of one of the above three dislocations. The reason why the reversible movement of partial dislocations during reverse transformation occurs by the addition of Si is not so clear at present. One possible explanation is the existence of internal stress.[44] A backward movement of the partial dislocations which operated in the forward transformation may be assisted by the back stress caused by the martensitic transformation. If plastic accommodation takes place in the austenite around the tip of the ε band, the back stress decreases. The solution hardening of austenite by Si may restrict the plastic accommodation. In the case of Fe–Mn binary alloys, the stress due to transformation strain is easily accommodated by local slip because of the low yield strength of austenite. In this sense, the solution hardening by Si is important for the reversible movement of partial dislocations. The second possible explanation is the existence of short range ordered structure or small coherent ordered particles in austenite, as was implied by Sade et al.[49] If the austenite has an ordered structure, the free energy of austenite is increased by an irreversible movement of partial dislocations and thus a reversible movement becomes preferential. However, ordered structure or ordered coherent particles in austenite have not been reported so far in Fe–Mn–Si alloys.

It must be emphasized that the high strength of austenite is essential both for the suppression of slip deformation during the stress-induced ε transformation and for the crystallographic reversibility during the $\varepsilon \rightarrow \gamma$ reverse transformation, and thus for the perfect shape memory effect. Therefore, for the further development of Fe–Mn based shape memory alloys, the strengthening of

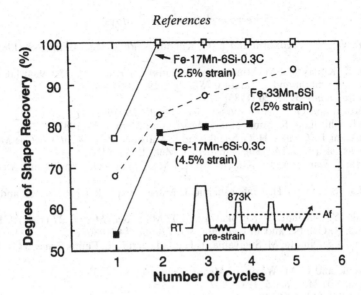

Fig. 5.11. Change in the degree of shape recovery with increase in the training cycles in Fe–17%Mn–6%Si–0.3%C and Fe–33%Mn–6%Si alloys. (After Tsuzaki et al.[18])

austenite must be taken into account. There have been some attempts to obtain a good shape memory effect in Fe–Mn–Co alloys, since Co also reduces the Néel point and lowers the stacking fault energy of austenite in a similar way as Si. However, Fe–Mn–Co alloys exhibit only a slight shape memory effect like Fe–Mn binary alloys.[50,51] This is probably due to the fact that Co does not strengthen the austenite matrix.

Interstitial atoms such as C and N strengthen the austenite of Fe–high Mn alloys. Therefore, the addition of C is expected to improve the shape memory effect. Actually, Tsuzaki et al.[18] have found that the shape memory effect is improved by the addition of carbon as shown in Fig. 5.11. This result suggests that the addition of other alloying elements strengthening the matrix also improves the shape memory effect associated with ε martensite.

References

1. T. Maki, *Proc. of Int. Conf. on Martensitic Transformations* (ICOMAT-89), ed. B. C. Muddle (Trans Tech Publication, 1989) p. 157.
2. R. G. Davies and C. L. Magee, *Metall. Trans.*, **2** (1971) 1939.
3. G. Thomas, *Metall. Trans.*, **2** (1971) 2373.
4. G. Krauss and A. R. Marder, *Metall. Trans.*, **2** (1971) 2343.
5. T. Maki, S. Shimooka, M. Umemoto and I. Tamura, *Trans. JIM*, **13** (1972) 400.
6. M. Umemoto, E. Yoshitake and I. Tamura, *J. Mater. Sci.*, **18** (1983) 2893.
7. C. M. Wayman, *Scr. Metall.*, **5** (1971) 489.
8. M. Foos, C. Frantz and M. Gantois, *Shape Memory Effect in Alloys*, ed. J. Perkins (Plenum Press, New York, 1975) p. 407.

9. Yu. N. Koval, V. V. Kokorin and L. G. Khandros, *Phys. Met. Metall.*, **48** (1981), No. 6, 162.
10. T. Maki, K. Kobayashi, M. Minato and I. Tamura, *Scr. Metall.*, **18** (1984) 1105.
11. T. Maki, S. Furutani and I. Tamura, *ISIJ Int.*, **29** (1989) 438.
12. S. Kajiwara, *Trans. JIM*, **26** (1985) 595.
13. Yu. N. Koval and G. E. Monastyrsky, *Scr. Metall.*, **28** (1993) 41.
14. A. Sato, E. Chishima, K. Soma and T. Mori, *Acta Metall.*, **30** (1982) 1177.
15. M. Murakami, H. Otsuka, H. G. Suzuki and S. Matsuda, *Proc. of Int. Conf. on Martensitic Transformations* (ICOMAT-86), (Japan Inst. Met., 1986) p. 985.
16. Y. Moriya, T. Sampei and I. Kozasu, Annual Meeting at Yokohama, Conf. Abstr., *JIM*, (1989) p. 222.
17. H. Otsuka, H. Yamada, H. Tanahashi and T. Maruyama, in Ref. [16], p. 222, and Ref. [1], p. 655.
18. K. Tsuzaki, Y. Natsume, Y. Kurokawa and T. Maki, *Scr. Metall.*, **27** (1992) 471.
19. T. Somura, R. Oshima and F. E. Fujita, *Scr. Metall.*, **14** (1980) 855.
20. R. Oshima, S. Sugimoto, M. Sugiyama, T. Hamada and F. E. Fujita, *Trans. JIM*, **26** (1985) 523.
21. D. P. Dunne and C. M. Wayman, *Metall. Trans.*, **4** (1973) 137.
22. R. Oshima, *Scr. Metall.*, **15** (1981) 829.
23. N. Jost, in Ref. [1], p. 667.
24. A. Sato, Y. Yamaji and T. Mori, *Acta Metall.*, **34** (1986) 287.
25. S. Kajiwara and T. Kikuchi, *Philos. Mag. A*, **48** (1983) 509.
26. T. Maki and C. M. Wayman, *Proc. of 1st JIM Int. Symp. on New Aspects of Martensitic Transformation* (JIMIS-1) (Japan Inst. Met., 1976) p. 75.
27. T. Maki and I. Tamura, in Ref. [15], p. 963.
28. S. Kajiwara and W. S. Owen, *Scr. Metall.*, **11** (1977) 137.
29. M. Umemoto and C. M. Wayman, *Metall. Trans. A*, **9A** (1978) 891.
30. S. Kajiwara and W. S. Owen, *Metall. Trans.*, **5** (1974) 2047.
31. T. Maki and C. M. Wayman, in Ref. [26], p. 69.
32. T. Maki and C. M. Wayman, *Metall. Trans. A*, **7A** (1976) 1511.
33. M. Umemoto and C. M. Wayman, *Trans. JIM*, **19** (1978) 281.
34. T. Maki, *Proc. of MRS Int. Mtg. on Advanced Materials*, vol. 9, ed. M. Doyama *et al.* (MRS, Pittsburgh, 1989) p. 415.
35. E. Hornbogen and W. M. Mayer, *Acta Metall.*, **15** (1967) 584.
36. J. K. Abraham and J. S. Pascover, *Trans. AIME*, **245** (1969) 759.
37. T. Maki and C. M. Wayman, *Acta Metall.*, **25** (1977) 695.
38. C. Magee and R. G. Davies, *Acta Metall.*, **20** (1972) 1031.
39. R. F. Decker, *Proc. of Symp. on Steel-Strengthening Mechanisms* (Climax Molybdenum Co., Zurich, 1969) p. 147.
40. H. Otsuka, M. Murakami and S. Matsuda, in Ref. [34], p. 451.
41. P. Gaunt and J. W. Christian, *Acta Metall.*, **7** (1959) 529.
42. A. Sato, E. Chishima, Y. Yamaji and T. Mori, *Acta Metall.*, **32** (1984) 539.
43. J. H. Yang and C. M. Wayman, *Mater. Characterization*, **28** (1992) 23 and 37.
44. T. Maki and K. Tsuzaki, *Proc. of Int. Conf. on Martensitic Transformations* (ICOMAT-92), ed. C. M. Wayman and J. Perkins (Monterey Institute for Advanced Studies, 1992) p. 1151.
45. A. Sato, H. Kasuga and T. Mori, *Proc. of Int. Conf. on Martensitic Transformations* (ICOMAT-79), (1979) p. 183.
46. G. Ghosh, Y. Vanderveken, J. Van Humbeeck, M. Chandrasekaran, L. Delaey and W. Vanmoorleghem, in Ref. [34], p. 457.
47. K. Tsuzaki, M. Ikegami, Y. Tomota, Y. Kurokawa, W. Nakagawara and T. Maki, *Mat. Trans. JIM*, **33** (1992) 263.
48. Y. Tomota, W. Nakagawara, K. Tsuzaki and T. Maki, *Scr. Metall.*, **26** (1992) 1571.
49. M. Sade, K. Halter and E. Hornbogen, *Z. Metallkd.* **79** (1988) 487.
50. C. F. Chen, J. H. Yang and L. C. Zhao, in Ref. [34], p. 481.
51. A. A. H. Hamers and C. M. Wayman, *Scr. Metall.*, **25** (1991) 2723.

6

Fabrication of shape memory alloys

Y. SUZUKI

Shape memory alloys (SMAs) in practical use are limited to three kinds: Ti–Ni alloys, Cu-based alloys and Fe-based alloys. Figure 6.1 shows typical fabrication processes of the first two types. Some problems exist in the fabrication of commercial SMAs: (1) control of the alloy composition; (2) cold-working; and (3) shape memory treatment.[1-3] (1) The Ti–Ni alloy is an equiatomic intermetallic compound and a composition shift from stoichiometry greatly affects the characteristics of the alloy. In particular, transformation temperature is extremely sensitive to composition. One percent shift in Ni content results in a 100 K change in the M_s or A_f point. While the Cu–Zn–Al alloy is not an intermetallic compound, the composition dependence of the transformation temperature is also large. One percent shift in the Al or Zn content changes the M_s temperature by 100 K or 60 K respectively. (2) Although the Ti–Ni alloy is known as a unique intermetallic compound which is workable in the cold state, actual cold-working is substantially harder. Cold-working of the Cu–Zn–Al alloy is also difficult, but the origin of the difficulty is different for both. The shape memory effect (SME) of the Cu–Zn–Al alloy appears when the alloy is in the β phase region. However, cold-working of the β phase alloy is impractical. Ingenious heat-treatment is employed to overcome this problem. The alloy is temporarily transformed into a workable phase and cold-worked. The worked product is transformed to the β phase in the final heat-treatment. (3) The cold-worked primary product such as a wire or a sheet is formed into the final product shape such as a helical spring. But the alloy product does not exhibit SME in the as-formed condition. A special heat-treatment, the so-called 'shape memory treatment', is necessary to reveal the desired SME. In the case of the Ti–Ni alloys, the formed product is fixed onto a jig and heated in an electric furnace. The heat-treated alloy obtains the shape memory property to recover to the memorized shape. The shape memory treatment of Cu–Zn–Al is slightly complex and will be mentioned later.[4]

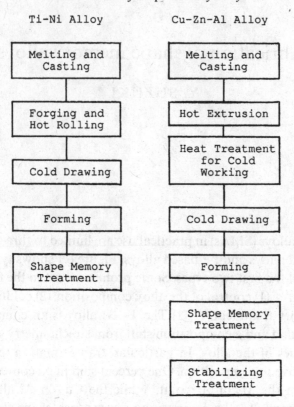

Fig. 6.1. Fabrication processes of Ti–Ni and Cu–Zn–Al shape memory alloys.

In this chapter, fundamental fabrication processes of Ti–Ni and Cu–Zn–Al alloys and some sophisticated techniques are presented.

6.1 Fabrication of Ti–Ni based alloys

6.1.1 Melting and casting processes

Since molten Ti is very reactive to oxygen, Ti–Ni alloy is melted in high vacuum or an inert gas atmosphere. A high frequency induction melting method is most commonly used. Electron beam melting, argon arc melting and plasma arc melting are also employed.[5] The first advantage of induction melting is the homogeneity of chemical composition throughout the ingot, since alternating current induction has a mixing effect on the molten alloy. The recommended crucible material is graphite or calcia (CaO). Alumina or magnesia are not suitable because the oxygen contained in the crucible contaminates the molten alloy. In the case of a graphite crucible, oxygen contamination is negligible, but carbon must be considered. Carbon content in the molten Ti–Ni alloy largely depends on the melt temperature. If the melt temperature exceeds 1723 K, the use of a graphite crucible is impratical. Fortunately, the melting point of the stoichiometric Ti–Ni alloy is *ca* 1510 K so that the melting procedure can be carried out at relatively low temperatures. The carbon content in the ingot prepared under a pertinent operation lies between 200 and 500 ppm. Such a small amount of carbon does not affect the shape memory characteristics of the alloy. Another advantage of induction melting is the controllability of the chemical composition. If the operation proceeds carefully, the M_s temperature of the ingot can be controlled to within ± 5 K. To achieve more precise control, an in-situ composition control furnace has been developed.[6] Figures 6.2 and 6.3 show the appearance of the

Fig. 6.2. Installation of an in-situ composition control furnace for Ti–Ni shape memory alloys. (After Suzuki *et al.*[6])

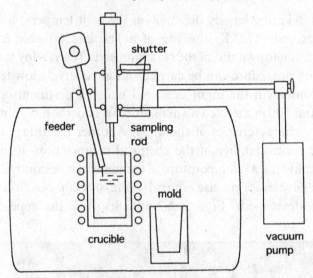

Fig. 6.3. Operation system of in-situ composition control furnace.

furnace and a schematic diagram of the operating system, respectively. The molten alloy is picked up from the melt in vacuum and the A_f point of the sample is measured quickly and then an alloying element is added to adjust the A_f point.

Electron beam melting utilizes a high-voltage electron beam as the heating source. Raw metals are melted by electron beam irradiation and drop down in a puddle in the water-cooled copper mold. The alloy is solidified from the bottom side and is pulled down through the bottom of the mold. The lowest impurity content is realized in this method because of the purifying effect of the high vacuum and high heating temperature. On the contrary, composition homogeneity in the ingot is insufficient because the alloy solidifies unidirectionally from the bottom. Evaporation of metal due to the high heating temperature complicates composition control. In spite of these shortcomings, this method is used to prepare Ti–Ni SMAs which do not require precise control of the transformation temperature.

The argon arc melting method is classified into two types with regard to the heating system. The first type uses a non-consumable electrode and the second a consumable electrode consisting of materials to be melted. The first method is preferred in laboratories because it is applicable to many kinds of alloys. In this method, raw metals are installed on a copper mold and irradiated by the argon arc from an electrode made of a tungsten rod. When the alloy is melted down, its shape resembles a button due to the surface tension effect. The solidified button shaped ingot is turned over and remelted repeatedly to

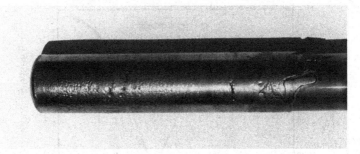

Fig. 6.4. Cylindrical ingot of a Ti–Ni alloy prepared by the plasma melting method. (After Suzuki *et al.*[6])

improve the homogeneity of the composition. In the second method, a furnace installs a consumable electrode which consists of raw materials. The electrode has two roles: a heating source and a material source. The electrode is heated by the argon arc and the molten alloy drops down onto the mold and forms a cylindrical ingot. The productivity of the second method is higher than the first.

The plasma melting method uses a low velocity electron beam discharged from a hollow plasma cathode.[7] Compared to the hard irradiation from a high voltage electron beam or an argon arc, electron irradiation from the plasma cathode is milder. It results in low loss of the alloy element. Composition distribution in the ingot is uniform in spite of the use of a water-cooled mold. Figure 6.4 shows a cylindrical ingot prepared by the plasma melting method.

6.1.2 Hot- and cold-working

After removing the surface layer, the ingot is forged and rolled into a bar or a slab with appropriate size. Rod or wire products are roll-worked using a bar rolling mill which has a pair of grooved rollers. Figure 6.5 shows the tensile strength and elongation of a Ti–Ni alloy at high temperatures. The tensile strength begins to decrease at 600 K and the decline accelerates above 650 K. The elongation rises at 800 K and exceeds 100% at 900 K. Thus the alloy is easily worked if the temperature is above 800 K. Although the workability of the Ti–Ni alloy is improved at higher temperatures, the alloy surface is more roughened by oxidation. The optimum heating temperature for hot-working is around 1073 K.

Compared to the hot-working, cold-working of Ti–Ni alloy is far more difficult. The workability strongly depends on the alloy composition. It becomes harder with increasing Ni content. Especially, working becomes difficult when the Ni content exceeds 51 at%. A source of the hard workability is strong

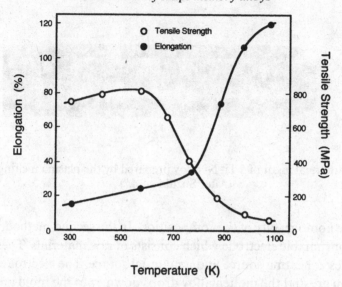

Fig. 6.5. Tensile strength and elongation of a 50.0 at% Ti–Ni alloy at high temperature.

work-hardening. Yield strength of annealed Ti–Ni alloy is lower than 100 MPa, which is as soft as annealed copper or aluminum. When the annealed Ti–Ni alloy wire is deformed, tensile stress rises at *ca* 10% strain and it reaches 1000 MPa at 40% strain. In the case of cold-drawing, the tensile strength surpasses 1500 MPa.

To carry out a satisfactory working, an optimal setting which combines the drawing and annealing processes has to be planned. Adhesion of wire to the tools often occurs so that selection of tools and lubricants is also of concern. An oxide layer on the wire surface improves lubrication against the tool. However, a thick oxide layer impairs the shape recovery characteristics.

Machining of Ti–Ni alloy is hard. Drilling is particularly difficult and tapping by conventional tools is impossible. Tungsten carbide tools are generally recommended because of a high tool life. There exists an optimum rate of machining which promotes long tool life, so the machining conditions should be determined by trials in advance of commercial production. Grinding of Ti–Ni is not difficult, but the abrasion loss of the grinding tool is high. GC (green carborundum) is recommended as a tool material.

6.1.3 Forming and shape memory treatment

The cold-drawn Ti–Ni wire is formed to the final product shape. A small number of springs are occasionally prepared by a simple method in which the

Fig. 6.6. Front view of an automatic coil forming machine. (Courtesy of Asahi Seiki Manufacturing Co., Ltd)

Ti–Ni wire is wound on a cylindrical jig. In most commercial production, the coil springs are formed with use of an automatic forming machine (Fig. 6.6). Figure 6.7 shows the forming process of the helical coil schematically. An SMA wire is driven by a pair of rollers B, B' and fed into a bending section through a guide C. During the time the wire slides on the coiling pins D, D' and an arbor E, it is formed into a helical coil. A pitch spacer F moves along the spring axis to set up the spring pitch. When the active turns of the spring reach a preset value, a cutter G cuts off the formed coil. The spring-back effect of Ti–Ni wire is larger than that of steel wire so that a larger forming amplitude has to be applied to get adequate forming.

The final process of Ti–Ni SMA fabrication is shape memory treatment. The most widely used treatment is the so-called 'medium temperature treatment.'[8] The actual procedure of this treatment is quite simple. The formed spring is fastened on a jig and then heated at 623–723 K to memorize the shape. Fastening is necessary not to change the spring shape during the heat-treatment. The holding time ranges from 10 to 100 min depending on the product

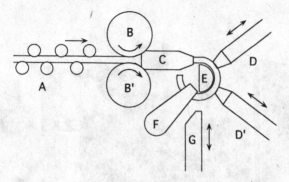

Fig. 6.7. Operating principle of the coil forming machine.

size. The most important point for this treatment is that the alloy wire is work-hardened adequately in advance of the forming. If the tensile strength of the wire is less than 1000 MPa, sufficient SME does not appear after the treatment. The tensile strength of the cold-drawn wire is checked prior to the forming process.

The heat-treatment temperature is adjusted according to detailed product specifications. If it is required to increase the difference between yield stresses at high and low temperatures such as for a bias type two-way actuator, the treatment temperature is set at higher than 673 K. On the other hand, if it is only required to increase the recovering force above the A_f point, a lower temperature treatment below 673 K is favored. Since the treatment temperature affects the transformation temperature and other shape memory characteristics, the furnace temperature is controlled precisely and the air in the furnace is circulated sufficiently to homogenize the temperature distribution. The air circulation is also effective in accelerating the heating rate of the product. When heating is accomplished, the alloy products are taken out and cooled.

The SME due to R-phase transformation has low thermal hysteresis and exhibits excellent fatigue properties.[9] A couple of methods are known to reveal the R-phase transformation: (1) addition of a certain alloying element such as iron; and (2) heating at medium temperature after strong work-hardening. The latter method is the same as the medium temperature treatment. It is noted that the alloy treated at a medium temperature exhibits both the R-phase and martensitic transformations at different temperatures.

In addition to the medium temperature treatment, several other kinds of shape memory treatments have been developed. The low temperature treatment is a dated method in which the alloy is normalized at 1073–1273 K and formed to the desired shape at room temperature and then heated at

473–573 K for 1–2 hours. The shape memory properties after this treatment are inferior to those of the medium temperature treatment. Another method is an aging treatment which is applicable only to high nickel alloys containing more than 50.5 at% Ni.[10] The alloy is solution-treated at high temperature and aged at *ca* 673 K for 1–5 hours to take a precipitation hardening. The aged alloy exhibits good SME comparable to an alloy after the medium temperature treatment. Aging treatment is also applied to reveal an all-round SME (see Chapter 3).[11] A Ti–Ni alloy sheet is solution-treated at 1073 K and fixed to an arch shape. The all-round SME appears after 100 hours heating at 673 K.

Heat-treatment of a superelastic alloy is almost the same as that of SMA, since SME and superelasticity both appear in the same alloy, and the only difference between the two lies in the transformation temperature. The medium temperature treatment is mostly used to reveal superelasticity and the aging treatment is sometimes used.

6.1.4 Finishing and testing processes

The corrosion resistance of Ti–Ni alloy is superior and surface treatment for corrosion protection is not needed. On the other hand, precious metal plating to beautify the surface appearance is popular in ornamental products. Since the plated layer does not possess SME, a thick layer undermines the shape recovery property. Optimum thickness lies near 10 μm. Electro-plating and ion-plating are both used in coating a gold layer on eye glass frames. Other surface coatings are also employed to improve the slip smoothness against mated materials. A nylon coating on bra supporting wires is for fitting with the fabric. The surface of catheter wires for medical use is coated with silicone resin.

Transformation temperatures of the Ti–Ni alloy are determined with the use of a differential scanning calorimeter (DSC). Figure 6.8 shows an example of a DSC chart. The upper line indicates the heat change in the specimen on heating and the lower line that on cooling. The transformation temperatures are determined from the intersection of a base-line and the maximum gradient line of a lambda type curve. M_s, M_f, A_s and A_f points indicate the start and finish temperatures of the martensitic transformation on cooling and heating respectively. M_s' and M_f' are the start and finish points of the R-phase transformation. Shape recovery characteristics are the most important testing factors for commercial SMA products. The conventional tensile test is used to check the mechanical properties of SMA wires and plates. Two or more stress vs. strain curves are recorded at the temperatures above the A_f point and below the M_s point. The recovering stress of the alloy is estimated from the curve

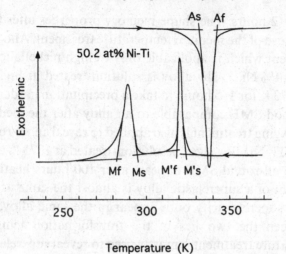

Fig. 6.8. Determination of transformation temperature from a DSC (Differential Scanning Calorimeter) chart.

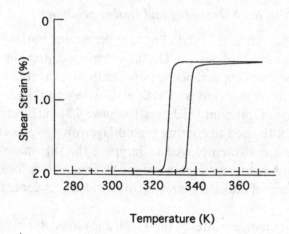

Fig. 6.9. Temperature vs. shear-strain curve of a Ti–Ni–Cu alloy spring under constant load.

above the A_f point and the deforming stress below the M_s point. Most widely referred data for an SMA spring is a temperature vs. deflection curve. The curve is provided by means of a temperature scanning method. An SMA spring is suspended under constant loading and its deflection is recorded continuously during temperature scanning. Figure 6.9 shows an example of temperature vs. deflection curves where the deflection is scaled with a shear-strain. A temperature vs. force curve is also used as a testing method. In this test, an SMA spring is deflected under a constant strain and the change in recovering force of a constrained SMA spring is recorded. The Japanese Standard

Association has issued several standards that specify testing items for SMA products.[12]

6.2 Fabrication of Cu–Zn–Al based alloys

6.2.1 Melting and casting processes

The melting process of Cu–Zn–Al alloys is fundamentally the same as the conventional process for commercial brass.[13] However, the required precision of the alloy composition is far more strict and the conventional processes cannot satisfy these requirements. There exist two problems in composition control: zinc oxidation and aluminum segregation. Zinc in the melts is oxidized and flies up from the surface of the molten alloy. Aluminum segregates gravitationally to the upper side of the ingot. A core-less type low-frequency induction furnace which has a large mixing effect on the molten alloy is preferably employed. A high frequency induction furnace is sometimes used to prepare a small size ingot. A fuel heating or electric heating furnace is not used because of a low heating rate and large oxidation loss. While a flux is not necessary to prepare the Cu–Zn based alloys, a compound flux composed of NaCl, KCl, etc. is often added to improve the surface cleanliness of the melt.

6.2.2 Hot- and cold-working

Hot-working of Cu–Zn–Al alloys, such as forging, rolling and extrusion, is easily carried out. Hot-extrusion is particularly efficient because of its large reduction ratio at one time. A billet of 200 mm in diameter can be directly extruded to a wire of 3 mm diameter. Accordingly, most of the primary products are extruded as bars with arbitrary cross-sections (round, rectangular, etc.).

The hot-worked alloy has a coarse grain structure which causes grain boundary rupture and depresses fatigue life of the alloy. A fine grain structure which improves grain boundary strength and fatigue life is one of the indispensable properties for commercial Cu–Zn–Al alloys. Cold-working is a most effective way to refine the grain size. According to the phase diagram, a Cu–Zn–Al alloy after hot-working is in the β phase but cold-working of the β phase alloys is impractical. In order to obtain cold-workability, the alloy is heat-treated to be transformed into a dual phase state where a small amount of the β phase is mixed in the workable α phase. Heat-treatment at 773–873 K provides such transformation. Cold-working up to 30% working-ratio is thereby realized.

The cold-worked dual phase alloy is then formed to the shape of the final product. In the case of Cu–Zn–Al alloys, a compression spring is the most used. Typical spring dimensions are 2–3 mm in wire diameter and 10–30 mm in coil diameter. The wire forming procedure is almost the same as that of the Ti–Ni alloy. There are no problems with adhesion of the forming tool and spring-back of the wire which occur in Ti–Ni alloy forming. The machinability of Cu–Zn–Al alloys is good and high-speed steel tools can be used without any trouble.

6.2.3 Heat-treatment and finishing processes

There are two objectives in the shape memory treatment of Cu–Zn–Al alloy: (1) memorizing a desired shape; and (2) transforming the alloy into the β phase. In order to satisfy both requirements, the alloy products are heat-treated at a temperature within the β phase region. The heating temperature is chosen to be as low as possible to avoid grain coarsening. After a short-time holding, the alloy is quenched in water. Cooling agents such as KOH are commonly used to accelerate the cooling rate. High-rate cooling is required to prevent α phase precipitation. The precipitation behavior of the α phase is inferred from isothermal transformation diagrams. Figure 6.10 shows a diagram for a Cu–20Zn–5Al alloy in which a precipitation nose appears at *ca* 673 K. The cooling curve has to slope down before the nose of the isothermal transformation curve. The alloy in as-quenched condition is thermally unstable and the transformation temperature shifts with aging at ambient temperature. The alloy after quenching is immediately heated at *ca.* 373 K and held for several tens of minutes to stabilize the intrinsic thermal state.

In applications of Cu–Zn–Al alloys, a two-way SME is often sought. Several kinds of techniques are known to induce a two-way SME: (1) large deformation beyond the recoverable limit; (2) heating under a constrained condition; and (3) repetition of deforming and heating. The most practical method is (3), the so-called 'training method'.[14] An alloy is deformed above the A_f point and fixed and then cooled down below the M_s point under constraint. The alloy is released after cooling and heated again to recover the original shape. A two-way SME appears after 10–20 repetitions of this thermal cycle. The recoverable strain of the two-way SME by training is roughly half of the deformed strain above the A_f point.

Since the Cu–Zn–Al alloy is sensitive to stress corrosion cracking (SCC), surface protection is required if the environmental conditions enhance the occurrence of SCC. Conventional tin plating provides sufficient protection against SCC. The optimum thickness of the plating layer is *ca* 15 μm. A thicker

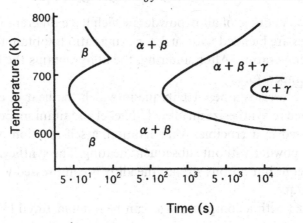

Fig. 6.10. Isothermal transformation diagram of a Cu–20Zn–5Al alloy. (After Delaey *et al.*[2])

plating layer impairs the SME but a thinner layer does not provide good protection.

6.3 Powder metallurgy and miscellaneous methods

Sophisticated fabrication techniques for SMAs such as powder metallurgy are now being developed. In the case of Ti–Ni alloys, the objective is near-net shape processing which enables an approximate finish to a final shape. In the case of Cu–Zn–Al alloys, a grain refinement effect to improve the fatigue properties is targeted.

6.3.1 Powder metallurgy and other techniques of Ti–Ni alloys

There are two kinds of powder metallurgy techniques for Ti–Ni alloys: raw metal powder sintering and alloy powder sintering. In the former method, pure metal powders are blended, pressed and sintered so that inhomogeneity in the alloy composition is inevitable. The latter method uses a pre-alloyed powder and homogeneity of the sintered alloy is further improved. The plasma rotating-electrode method provides a high quality alloy powder for sintering. An electrode consisting of the alloy which is heated by a plasma arc rotates with high speed. Droplets of the molten alloy are thrown centrifugally and atomized. The solidified powder is sintered and HIP (Hot Isostatic Press) processed. The processed alloy exhibits a good SME that is comparable to the vacuum melted alloy.[15]

Since powder metallurgy does not involve a melting process that causes composition inaccuracy, it can be applied to precise control of the transform-

ation temperature. A couple of alloy powders which have different transformation temperatures are blended with such a mixing ratio to obtain the desired
transformation temperature. After sintering, the alloy obtains the calculated
transformation temperature.[16]

Combustion synthesis is a new technique in which the heat of compound
formation is utilized to synthesize an alloy.[17] Nickel and titanium powders are
blended and pressed in a crucible. After ignition, a self-reaction propagates
through the alloy powder without subsequent heating. The synthesized Ti–Ni
alloy is homogeneous but porous so that the alloy is compressed with the use
of the HIP technique.

An alloy product with a complex shape can be manufactured by precision
casting techniques. A pressure casting method is used as follows to cast dental
components such as a crown or a clasp.[18] A mold is prepared by conventional
dental techniques. A small Ti–Ni block is set on the top of the mold and heated
by an argon arc. When the alloy block is melting down, molten alloy is
squeezed into the mold with the aid of argon gas pressure. After the alloy has
solidified, the mold is broken to take out the alloy cast. The alloy cast is then
heat-treated at 1023 K. The shape recoverability of the cast alloy is roughly
half that of the cold-worked alloy.

The rapid quenching method, in which a molten alloy ejected from a nozzle
is thrown onto a rotating drum, was originally developed to prepare amorphous metals. A thin Ti–Ni ribbon can be prepared with the use of this
technique.[19] In particular, an alloy with a new composition which is not
realized in the equilibrium state can be prepared. Ti–Ni–Cu alloys whose Cu
content exceeds 10% exhibit a unique property which is not shown in the
conventionally processed alloys.

A Ti–Ni film thinner than 10 μm is prepared by means of vapor phase
deposition.[20] Sputtering is the most practical method. The evaporation or
ion-plating method does not suit Ti–Ni alloys because of the lack of controllability of chemical compositions. The principle of film growth by the sputtering
method is illustrated in Fig. 6.11. A target electrode consisting of a Ti–Ni alloy
and a substrate are positioned opposite one another. In the case of Ti–Ni films,
glass or a rock-salt is used as a substrate. Argon gas is supplied between them
and is ionized. Argon ions accelerated by an electric field bombard the target
and strike out (sputter) the alloy atoms. The sputtered atoms deposit on the
opposite substrate.

6.3.2 Powder metallurgy of Cu–Zn–Al alloys

Practical Cu–Zn–Al alloys contain small amounts of Ti, Zr or Nb for the

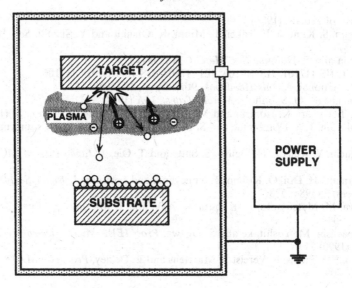

Fig. 6.11. Schematic diagram of sputter film growth process.

purpose of grain refinement (with regard to grain refinement, see Chapter 4). The grain refining effect of the additive elements is, however, not satisfactory and the grains are easily coarsened during subsequent heat-treatment. Powder metallurgy is an effective way to obtain a stable fine-grain structure.[21] Cu–Zn–Al alloy powder is prepared by a water atomizing method which provides a cheap alloy powder. The alloy powder is pre-pressed to a billet shape and extruded at *ca* 1073 K. The extruded bar is immediately quenched in water. The grain sizes of the as-quenched alloy are *ca* 30 μm. The processed alloy contains a number of fine, elongated alumina particles which disperse homogeneously throughout the extruded bar. Owing to this microstructure, the fatigue properties of the alloy are improved significantly.

References

1. W. J. Buehler and W. B. Cross, *Wire Journal*, (June 1969) 41.
2. L. Delaey, A. Deruyttere, E. Aernoudt and J. R. Roos, *INCRA Report* 78R1, (1978).
3. Y. Suzuki, *Jitsuyo Keijo-Kioku-Gokin* (in Japanese), (Kogyo Chosa-kai, 1987).
4. H. Funakubo, ed., *Shape Memory Alloys*, (Gordon & Breach Sci. Pub., 1987).
5. R. F. Bunshah, ed., *Techniques of Metals Research*, Vol. 1, part 2, (Interscience Publishers, 1968).
6. Y. Suzuki, *MRS Int'l Mtg. on Adv. Mats.*, Vol. 9, (1989) 557.
7. Y. Suzuki, *Bulletin of the Japan Inst. Metals* (in Japanese), **24** (1985) 41.
8. Y. Suzuki, see Ref. [4], p. 194.
9. H. Tamura, Y. Suzuki and T. Todoroki, *Proc. of Int. Conf. on Martensitic Transformation*,

Japan Inst. of Metals, (1986) 736.
10. S. Miyazaki, S. Kimura, F. Takei, T. Miura, K. Otsuka and Y. Suzuki, *Scr. Metall.*, **17** (1983) 1057.
11. M. Nishida and T. Homma, *Scr. Metall.*, **18** (1984) 1293.
12. JIS H7001, JIS H7101, JIS H7103, JIS H7104, JIS H7105, JIS H7106.
13. ASM International, *Metals Handbook*, 9th edition, **15** (1988) 771.
14. J. Perkins and R. O. Sponholz, *Metall. Trans.*, **15A** (1984) 313.
15. H. Kato, T. Koyari, K. Isonishi and M. Tokizane, *Scr. Metall. et Mater.*, **24** (1990) 2335.
16. W. A. Johnson, J. A. Dominique and S. H. Reichman, *J. de Phys.*, **C4** supplement, (1982) C4-285.
17. M. Otaguchi, Y. Kaieda, N. Oguro, S. Shite and T. Oie, *J. Japan Inst. Metals*, **54** (1990) 214.
18. H. Hamanaka, H. Doi, O. Kohno, T. Yoneyama and I. Miura, *J. Japan. Soc. Dental Mat. and Devices*, **5** (1986) 578.
19. Y. Furuya, M. Matsumoto, H. Kimura, K. Aoki and T. Masumoto, *Mater. Trans. JIM.*, **32** (1990) 504.
20. K. Kuribayashi, M. Yoshitake and S. Ogawa, *Proc. IEEE Micro Electro Mechanical Systems*, (1990) 217.
21. J. Janssen, F. Willems, B. Verelst, J. Maertens and L. Delaey, *Proc. ICOMAT82*, Belgium, (1982) C4-809.

7

Characteristics of shape memory alloys

J. VAN HUMBEECK AND R. STALMANS

Symbols

a parent phase

A_f reverse transformation finish temperature

A_s reverse transformation start temperature

$\sum A$ summation on all martensite–austenite interfaces dA

d_a crystal defect in the austenitic structure

e_{ir}(m–a) thermodynamic irreversible contributions during transformation

E modulus of elasticity

f_M the average of the function f on the mass M_0

$f_{\Sigma A}$ the average of the function f on all austenite–martensite interfaces dA

Δf change of the function f during transformation or reorientation

g_{ch} chemical free energy density

g_d local defect energy density

g_{eM} average elastic energy density

$\Delta g_{d\Sigma A}([x])$ change of the average defect energy density on all austenite–martensite interfaces at a transformation fraction $[x]$

G_e total elastic free energy in the system

ΔK the stress intensity factor

m martensitic phase

m_i martensite variant i

M_f martensite finish temperature

M_0 total mass of the system

M_s martensite start temperature

Q^{-1} loss factor

r position

Δs entropy change during transformation from austenite to martensite

T temperature

T_c contact temperature in the case of generation of recovery stresses

$\hat{u}$ surface of unit area

x martensite mass fraction

$[x(r)]$ volume function which describes at any point r the phase and the crystallographic orientation of the phase

Z_g global Z-factor

Z_l local Z-factor

ε strain

ε_c contact strain in the case of generation of recovery stresses

ε_{ct} contact strain adapted for thermal dilatation

ε_e elastic strain

ε_t total strain

ε_{th} strain caused by thermal dilatation

ε_{tr} transformation or reorientation component of the total strain

$\Delta \varepsilon_{tr}$(m$_j$–a) transformation strain linked with transformation from austenite to the martensite variant m$_j$ resolved along the length axis

$\Delta \varepsilon_{tr\Sigma A}([x])$ average transformation strain on all austenite–martensite interfaces at a transformation fraction $[x]$

ρ local mass density

ρ_0 mass density in the unloaded condition

σ stress

σ_r recovery stress

σ_{rm} maximum recovery stress

σ_v yield stress

$d\sigma_r/dT$ stress rate during generation of recovery stresses

Ω free energy of the crystal defect d

7.1 Summary of the functional properties

In a general sense, shape memory behaviour refers to a peculiar thermo-mechanical behaviour in which small changes of one or two of the three

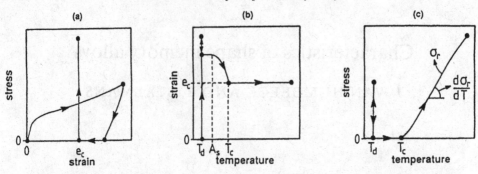

Fig. 7.1. The generation of recovery stresses is shown in three two-dimensional figures: (a) stress–strain, (b) strain–temperature, and (c) stress–temperature. A deformation is imposed at a temperature T_d in the martensitic state. Shape recovery is impeded at a contact strain e_c. From the corresponding contact temperature T_c recovery stresses σ_r are generated at a stress rate $d\sigma_r/dT$.

thermomechanical variables (temperature, strain and external stress) can induce extremely large changes of the two other thermomechanical variables. In this manner also, shape changes (e.g. reverse shape memory effect[1,2]) associated with preferentially oriented bainitic transformations are classified as shape memory behaviour. It is however clear that strictly speaking these effects have to be considered as shape effects, rather than memory effects. In a narrower sense, shape memory behaviour refers to the recovery of a large, apparently plastic deformation during heating: the so-called one-way memory effect. Mostly, and also in this chapter, only these shape memory properties are considered that are linked with a crystallographically reversible, thermoelastic transformation, as is the case for several metallic alloys. This reversible transformation of the parent phase to the product phase (martensite) is the basis of the following functional shape memory properties.

Below a critical temperature the material can be deformed to an imposed 'cold' martensitic shape. The apparently plastic deformation is recovered during subsequent heating to the parent phase. This effect is called the *one-way memory effect* since only the 'hot', parent shape is remembered (see Chapter 1). *Recovery stresses* are generated when the one-way memory effect is impeded. Since the recovery of the hot shape is impeded, a more specific name would be 'hot shape' recovery stresses (Fig. 7.1). The shape recovery during heating can also occur against an opposing force. This is the *work production during heating* or *actuator function* of shape memory alloys. The *two-way memory effect* (TWME) refers to the reversible (two-way) spontaneous shape change by heating and cooling between a hot parent shape and an 'acquired' cold martensitic shape. The foregoing shape memory properties require a temperature change, in contrast to the superelastic effect: isothermal loading of the hot shape results in large deformations at relatively low stresses; the hot shape is completely recovered during unloading (see Chapter 2). Recently evidence has been given for two new shape memory properties which are only present in samples showing the two-way memory effect. Firstly, it was shown that '*cold shape*' *recovery stresses* are generated when the shape change to the acquired 'cold' shape is impeded during cooling. Secondly, the shape change towards the cold shape during cooling can also occur against significant opposing forces. Hence, samples showing the TWME can do *work during cooling*.

These functional properties are closely related to the crystallographically reversible, thermoelastic, martensitic transformation. Because of this common basis, there are also several limits common to the different shape memory properties. Firstly, the macroscopic shape change is maximum when a single crystal of the most preferentially oriented martensite variant is formed. Making use of the crystallography of the transformation, this limit can be calculated from the orientation of the parent phase (see Chapter 1). Any additional strain generates irrecoverable plastic deformations with the result that the shape recovery is incomplete. The maximum recoverable strain in polycrystalline material is lower than in a single crystal of the same alloy due to (i) the different orientations of the grains in polycrystalline material and (ii) the strain compatibility requirements of adjacent grains. Secondly, recoverable shape changes require stresses lower than the critical stress for plastic deformation. In the case that the stresses are too high, irreversible plastic deformations occur and the result is an incomplete shape recovery. Thirdly, the parent phase and the martensitic phase are metastable phases. Consequently, overheating can result in the transformation to intermediate or stable phases and a loss of shape memory properties.

In this chapter follows a detailed discussion of the TWME (Section 7.3), the generation of recovery stresses (Section 7.4), the high damping capacity of shape memory alloys (Section 7.5) and the lifetime of shape memory alloys (Section 7.6). Property values are summarized in Section 7.7. First a generalized thermodynamic model which can be used to explain and predict the global shape memory behaviour is presented.

7.2 A generalized thermodynamic description of shape memory behaviour

7.2.1 Modelling the shape memory behaviour

Since the three-dimensional (stress σ – strain ε – temperature T) shape memory behaviour is determined by the thermomechanical history and path, it is impossible to measure the σ–ε–T relationship in all circumstances. Therefore, many mathematical and empirical models have been proposed to predict the thermomechanical behaviour of shape memory alloys.[3-6] However, these models show important shortcomings. The mathematical models are in general not very detailed: it is, for example, assumed that the transformation takes place at one definite temperature or stress, or the influence of hysteresis and thermomechanical history is not incorporated, etc. Moreover, the mathematical models contain a lot of coefficients which have to be determined experimentally. Empirical models are obtained by regression of a limited number of experimental data. However, as stated above, it is not sufficient to determine the σ–ε–T relationship for a set of constant stresses or temperatures to obtain a complete empirical model of the three-dimensional shape memory behaviour. As a result, both types of models are characterized by a very limited field of application, which has resulted repeatedly in erroneous extrapolations.

The thermoelastic, martensitic transformation has also been the subject of many thermodynamic studies.[7-27] However, these thermodynamic models also fail in many respects in the description of the thermomechanical behaviour of polycrystalline shape memory alloys, as the following examples illustrate.

(i) Many criteria have been formulated to describe the influence of external stresses on the selection of martensite variants.[7,14,27-32] These criteria can all be reduced to the criterion of maximum work.[14,33] This thermodynamic criterion states that the variants are formed which results in a maximization of the product of external stress and transformation strain, i.e. of the mechanical work done on the sample. However, it can be easily shown that the field of application of this criterion is rather limited. In the case of uniaxial loading this criterion predicts a complete transformation to a single martensite variant. In the case of parent single crystals this has been repeatedly experimentally confirmed.[30-31,34-37] On the other hand, a more complex behaviour has been observed in polycrystalline materials: in every parent grain several martensitic variants are formed during transformation of a uniaxially loaded sample, so that the criterion of maximum work fails in this more general case. It is also generally concluded from this criterion that the product of external stress and transformation strain is always positive during transformation. In contradiction to this, it has been shown recently that trained samples can elongate during cooling against an external compression stress.[38-39]

(ii) The influence of external stresses on the transformation temperatures has been described by different versions of the Clausius–Clapeyron equation.[9,22] From this equation, it has been inferred that external stresses always result in an increase of the M_s temperature. However, it has also been shown recently that external stresses opposing the TWME during cooling can result in a significant decrease of M_s.[38-39]

As will become clear in the following paragraphs, these thermodynamic theories of the thermoelastic martensitic transformation fail in the description of the thermomechanical behaviour because these theories are based on, many times implicitly, a defect-free parent phase and because the stored elastic energy is ignored or misrepresented. Therefore, taking into account the contribution of defects and the stored elastic energy, a generalized thermodynamic equilibrium equation, selection criterion and Clausius–Clapeyron equation will be formulated in the following paragraphs. Important examples of successful applications of this new thermodynamic description are the modelling of the two-way shape memory behaviour (Section 7.3) and of recovery stress generation (Section 7.4).

7.2.2 *The thermodynamic contribution of defects*

Consider a defect d_a with a free energy $\Omega_d(a)$ in the parent structure. During transformation this defect is 'inherited' by the martensitic structure. Since the crystallographic orientation of this defect depends on the orientation of the martensitic variant, the defect energy in the martensitic structure can be also variant-dependent. Therefore the defect energy is indicated by $\Omega_d(m_i)$, in which the subscript i refers to the martensitic variant m_i. Since the transformation and retransformation are diffusionless, retransformation results again in the original defect d_a. During transformation, this defect gives rise to a reversible free energy contribution $\{\Omega_d(m_i) - \Omega_d(a)\}$. The fact that $\Omega_d(m_i)$ can be different from $\Omega_d(m_j)$ also implies that *some* variants can become energetically favoured by the presence of this defect. With a view to the thermodynamic calculations in Section 7.2.4, the *local defect energy density* $g_{\bar{d}}$ is defined as:

$$g_d = (1/dm){*}\textstyle\sum_{dV}\Omega_d, \tag{7.1}$$

where dm is the mass of the infinitesimal volume dV, and $\sum_{dV}$ is a summation on all defects in the volume dV.

Defects which are of special interest in this respect are the complex dislocation arrays generated by cycling through the transformation region. High densities of these complex dislocation arrays with similar characteristics are produced by thermomechanical cycling[40–50] and by thermal cycling.[51–61] It is also found that there exists a one-to-one crystallographic correspondence between these dislocation arrays and the repeatedly induced martensite variants.[48] A thermodynamic analysis of specific experiments on Cu-based shape memory alloys[39] has evidenced that these dislocation arrays form low energy configurations in the repeatedly induced martensite variants, and high energy configurations in the other crystallographically equivalent variants. The resulting effect is that the thermodynamic contribution of these dislocation arrays favours the formation of the repeatedly induced variants.

7.2.3 *The thermodynamic contribution of the stored elastic energy*

The *stored 'elastic' energy* can be subdivided into several categories. *Elastic strain energy* is stored due to the external stresses and the corresponding elastic deformations. *Elastic strain energy* is also stored as a result of the elastic accommodation of the shape and volume changes during transformation. In the case of a thermoelastic martensitic transformation the interfaces are highly coherent, so that the *interface energy* can be neglected with respect to the elastic strain energy. The *average elastic energy density* is represented by g_{eM}:

$$g_{eM} = G_e/M_o, \tag{7.2}$$

where G_e is the total elastic free energy in the system, and M_o the total mass of the system.

7.2.4 *The thermodynamic description*

Since the existing thermodynamic models fail in many respects in the quantitative description of shape memory behaviour, the authors have developed a generalized thermodynamic model.[33] The approximations in this model have been selected in such a way that the model is a close approximation of the real system and that the final mathematical expressions are relatively simple.

A relatively simple system is considered[†] to avoid superfluous complications of the mathematical working out. This system is a uniaxially loaded bar with a homogeneous temperature T and free of macro residual stresses[‡]. Also, the small volume change during transformation is neglected. Irreversible contributions and the resulting hysteresis are neglected at the moment, but will be discussed in Section 7.2.6. Some notions and definitions are introduced here so as not to disturb the thermodynamic derivations.

(i) (z), (a), (m), and (m$_j$) refer to the phases z, parent, martensite, and the martensitic variant m$_j$, respectively.
(ii) x represents the martensite mass fraction, also indicated as transformation fraction. The volume function [x] gives in addition in any point r the phase and the crystallographic orientation of the phase; consequently, this function also describes the martensite variant arrangement.

[†] A generalized derivation without these restrictions can be performed along the same lines. However, the equations would become more complex with the result that the analysis becomes less clear.
[‡] Macro residual stresses can be tackled in a similar way as external stresses.

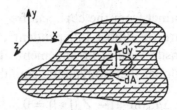

Fig. 7.2. An infinitesimal increase of the transformation fraction x occurs by the forward displacement dy of the crosshatched martensite–parent interface dA, situated in the xz-plane.

(iii) The average of the function f on the mass M_o is represented by f_M. The average of the function f on all parent–martensite interfaces dA (Fig. 7.2) is represented by $f_{\Sigma A}$, where ΣA represents a summation on all parent–martensite interfaces dA; dy represents an infinitesimal displacement perpendicular to dA, and ρ the local mass density:

$$f_{\Sigma A} = \frac{\displaystyle\int_{\Sigma A} f(z,r)*\rho*dy\,dA}{\displaystyle\int_{\Sigma A} \rho*dy\,dA}. \tag{7.3}$$

(iv) The change of the function f during transformation or reorientation is represented by Δf:

$$\Delta f(z_i - z_j) = f(z_i) - f(z_j). \tag{7.4}$$

So $\Delta\varepsilon_{tr}(m_j\text{–a})$ represents the transformation strain linked with transformation from the parent phase to the martensite variant m_j, resolved along the length axis.

Application of the first and second laws of thermodynamics results, after a lengthy calculation,[33] in a *global* thermodynamic equilibrium condition:

$$\Delta g_{ch}(\text{m–a}) + \Delta g_{d\Sigma A}([x]) + \partial g_{eM}/\partial x - (1/\rho_0)* \ \sigma*\Delta\varepsilon_{tr\Sigma A}([x]) = 0, \tag{7.5}$$

where $\Delta g_{d\Sigma A}([x])$ and $\Delta\varepsilon_{tr\Sigma A}([x])$ represent the change of the average defect energy density, and the average transformation strain respectively, on all parent–martensite interfaces at a transformation fraction $[x]$ (see Eq. (7.3)); ρ_0 is the mass density in the unloaded condition.

This equation can be simplified by introducing the global Z_g-factor:

$$Z_g([x]) = -\Delta g_{d\Sigma A}([x]) - \partial g_{eM}/\partial x + (1/\rho_0)*\sigma*\Delta\varepsilon_{tr\Sigma A}([x]). \tag{7.6}$$

This Z_g-factor consists of the three non-chemical free energy terms: the *defect term* $\{\Delta g_{d\Sigma A}([x])\}$, the *elastic term* $\{\partial g_{eM}/\partial x\}$, and the *work term* $\{(1/$

$\rho_0)^*\sigma^*\Delta\varepsilon_{tr\Sigma A}([x])\}$. As distinct from the chemical free energy, the composing terms of the Z_g-factor are (i) almost independent of temperature and (ii) dependent on the martensite variant arrangement. By introducing this Z_g-factor, the condition of global thermodynamic equilibrium is simplified to:

$$\Delta g_{ch}(m\text{--}a) - Z_g([x]) = 0. \tag{7.7}$$

It can also be deduced that the following local thermodynamic equilibrium condition holds at any point of a parent–martensite interface:

$$\Delta g_{ch}(m\text{--}a) - Z_1 = 0. \tag{7.8}$$

The local Z_1-factor at the position r is given by, with $\hat{u}$ a surface of unit area:

$$\begin{aligned}Z_1(m_j\text{--}a,r) = &- \Delta g_d(m_j\text{--}a,r) - (M_o/\rho^*\hat{u})^*(\partial g_{eM}/\partial y) \\ &+ (1/\rho_0)^*\sigma^*\Delta\varepsilon_{tr}(m_j\text{--}a,r).\end{aligned} \tag{7.9}$$

In Section 7.2.1 it was indicated that the criterion of maximum work is only valid in the case of uniaxial loading of a defect-free parent single crystal. Therefore, a generalized criterion is inferred from the equation expressing local thermodynamic equilibrium (Eq. (7.8)). Because of the three variant dependent terms, only one variant, represented by m_p, can fulfil this equation at the position r of the transformation front. For any fictitious variant m_k different from m_p it follows that:

$$Z_1(m_k\text{--}a) < Z_1(m_p\text{--}a). \tag{7.10}$$

In words: in any point of the transformation front, variants are formed which correspond to a maximum Z_1-factor.

The Z-factor is composed of a defect term, an elastic term and a work term. The influence of these three almost temperature independent terms on the selection of the martensite variants can be summarized as follows: (i) the defect term favours the formation of the variants which have been repeatedly induced during previous transformation cycles (see Section 7.2.2), (ii) the work term favours the formation of the variants showing a maximum transformational work, and (iii) the elastic term is minimum for a self-accommodating variant arrangement (see Section 7.2.5).

The stress and temperature conditions at which two phase equilibrium between martensite and the parent phase can exist, are in general described by means of Clausius–Clapeyron equations (Eq. (1.22)). However, there is in the literature some unclarity in the correct definition of the transformation strain used in the denominator of this equation. Therefore, a generalized Clausius–Clapeyron equation is discussed below. Differentiating Eq. (7.7) at constant transformation fraction [x] directly gives the stress and temperature conditions at which thermodynamic two-phase equilibrium is maintained:

$$d\sigma/dT = - \rho_0^* \Delta s/\Delta\varepsilon_{tr\Sigma A}([x]).$$ (7.11)

The condition of constant [x] also implies that the right-hand part of Eq. (7.11) is constant. Hence, Eq. (7.11) expresses a linear relationship between the transformation temperature $T([x])$ and the external stress σ. The condition of constant [x] is absolutely essential: from the moment that deviations of [x] appear, Eq. (7.11) cannot be applied and deviations of the linear relationship between σ and $T([x])$ can occur. The transformation strain in the right-hand part of Eq. (7.11) is not an alloy constant: $\Delta\varepsilon_{tr\Sigma A}([x])$ decreases in general with increasing x and is influenced also by other factors, including the heat treatment and the thermomechanical history.[62-65]

7.2.5 Some illustrations

It has been observed that uniaxial loading of a defect-free, parent single crystal results in the formation of a single martensite variant which corresponds to the criterion of maximum work.[30-31,34-37] This variant corresponds also to a maximum Z-factor since the Z-factor is in this case reduced to the work term:

$$Z = \sigma^* \Delta\varepsilon_{tr}(m_p - a)/\rho_0.$$ (7.12)

This example also evidences the limited field of application of the criterion of maximum work. Since only one variant is formed, $\Delta\varepsilon_{tr\Sigma A}([x])$ can be reduced to $\Delta\varepsilon_{tr}(m_p - a)$. Consequently, the generalized Clausius–Clapeyron equation is for this particular case reduced to the equation deduced by Wollants et al.[9,14,26]

The case of a polycrystalline parent sample during the first transformation cycle after annealing at sufficiently high temperatures (see Section 7.2.2) approximates the case of a defect-free system. The following has been observed: (i) a self-accommodating variant arrangement is formed in the absence of external stresses, and (ii) a constant external stress during transformation results in the formation of a preferentially oriented variant arrangement; the preferential character decreases with increasing transformation fraction x.[62,63]

The Z_g-factor is in the absence of external stresses reduced to the elastic term:

$$Z_g = - \partial g_{eM}/\partial x.$$ (7.13)

Consequently, maximization of the Z_g-factor implies that *at any time during transformation* the variants which result in a minimum increase of the elastic free energy are formed. This corresponds to a minimum stored elastic energy

$g_{eM}([x])$ at any transformation fraction $[x]$. This condition can only be fulfilled by a self-accommodating variant arrangement in agreement with experimental observations.

In the case that external stresses are applied, the local Z_1-factor becomes:

$$Z_1 = \sigma^*\Delta\varepsilon_{tr}(m_j-a)/\rho_0 - (M_o/\rho^*\hat{u})^*(\partial g_{eM}/\partial y). \tag{7.14}$$

The elastic term is equal to zero at the start of transformation, so:

$$Z_1(x = 0) = \sigma^*\Delta\varepsilon_{tr}(m_j-a)/\rho_0. \tag{7.15}$$

It follows that the transformation starts in the most favourably oriented parent grain by the growth of the variant m_{p1} corresponding to a maximum work term. Because of the shape change accompanying the growth of this preferential variant, the elastic term increases gradually with x, while the work term remains constant. Therefore, the growth of the variant m_{p1} will be followed by the growth of a variant m_{p2}. This variant m_{p2} is characterized by a smaller transformation strain (thus a smaller work term) and also a less increasing elastic term. This reasoning can be continued. The result is that gradually a more self-accommodating variant arrangement is formed in agreement with the experimental observations.

7.2.6 The influence of hysteresis on the thermoelastic equilibrium and thermomechanical behaviour

If thermodynamic reversible terms are the only contributions to the non-chemical free energy, a martensite plate can grow and disappear without any temperature or stress hysteresis. Olson et al.[66,67] have shown that the nucleation of thermoelastic martensite can induce only a very small hysteresis. A much larger hysteresis is mostly observed in practice. This hysteresis is the result of thermodynamic irreversible contributions, of which friction during the movement of the martensite–parent interfaces is by far the most important contribution.[20,68] Without any pronouncement on the nature of these irreversible contributions, the hysteresis can be *formally* described by adding to the equilibrium equation a term $e_{ir}(m-a)$ which expresses the irreversible contributions. Since this irreversible term hampers the transformation as well as the retransformation, this term must be positive during transformation and negative during retransformation. It is generally accepted that the irreversible term $e_{ir}(m-a)$ is almost temperature and stress independent, in which case the Z-factor criterion and the generalized Clausius–Clapeyron equation are still valid.

In the absence of irreversible contributions there exists a two-phase equilibrium *between the parent phase and martensite*. An important consequence of the irreversible term $e_{ir}(m-a)$ and the related hysteresis, is that the transformation direction of the two-phase equilibrium has to be specified. During transformation or retransformation, there exists a two-phase equilibrium *from the parent phase to martensite*, or *from martensite to the parent phase*, respectively. When a positive (negative) change of the independent variables σ or T is followed by a negative (positive) change, a hysteresis region has to be overcome in which there is neither transformation nor retransformation. The important influence of the irreversible term on the thermomechanical behaviour is further illustrated in Section 7.5.2.

7.3 Two-way memory behaviour

7.3.1 Some aspects of the two-way memory effect

The essential difference between the two-way memory effect (TWME) and the other shape memory properties is that in the case of the TWME the macroscopic shape change is generated spontaneously, i.e. without external stresses. In the literature it is generally accepted that the *origin* of this spontaneous preferential martensite formation has to be attributed to some sort of anisotropy in the substructure of the parent phase.[65,69] The *mechanism* of the spontaneous shape change requires an interaction between this substructural anisotropy and the martensite formation. This substructural anisotropy is not an inherent characteristic of the parent phase, but is only obtained after specific thermomechanical treatments, mostly indicated as training procedures (Section 7.3.2).

The exact origin and mechanism of the TWME obtained after training cycling can be easily explained by use of the thermodynamic description. The substructural anisotropy, i.e. the origin of the TWME, has to be attributed to the dislocation arrays which are generated during training cycling (Section 7.2.2). The defect energy of those dislocation arrays is minimum in the 'trained' variants, i.e. the preferentially oriented variants which have been repeatedly induced during training cycling. From the Z-factor criterion it follows that the growth of those 'trained' variants will be thermodynamically favoured, also during subsequent thermal cycling, thus explaining the TWME.[39]

The relaxation of inhomogeneous macro residual stresses is often accompanied by shape changes. A typical example is the warping of a piece of metalwork during annealing. Analogously, the TWME has been ascribed to the relaxation of residual stresses by preferential martensite forma-

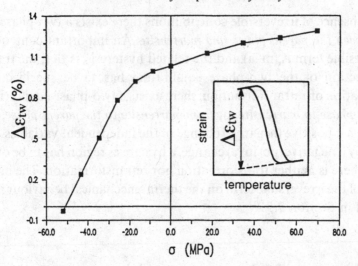

Fig. 7.3. The two-way memory strain $\Delta\varepsilon_{tw}$ as a functional of the external stress σ applied during cooling. (After Stalmans[39])

tion,[43,48,69-71] however without making the essential distinction between macro and micro residual stresses.[†] Inhomogeneous plastic deformation of either martensite or the parent phase results in an *inhomogeneous* residual macro stress distribution. This inhomogeneous residual macro stress distribution can, similarly to an external stress, result in preferential martensite formation and an accompanying shape change. From a physical point of view this reversible shape effect, which is attributed to a macro-scale asymmetry, is comparable with the one-way memory effect, but clearly distinct from the TWME which is attributed to a substructural anisotropy, i.e. a micro-scale asymmetry.

The most important TWME characteristic is its magnitude. It is also required that the magnitude of the TWME is highly reproducible, i.e. almost constant in successive thermal cycling. For a long time it has been believed that the TWME can be easily suppressed by applying a small opposing stress during cooling. In contrast, Fig. 7.3 shows that the two-way memory strain decreases *gradually* with increasing external opposing stress and that a relatively high opposing stress is required to suppress the TWME completely during cooling.[72] These observations can be easily explained by considering the influences of the three terms composing the Z-factor.[33] The stability of the TWME, i.e. the ability of the TWME to act against opposing stresses, can be considered as an additional TWME characteristic.

[†] Macro residual stresses, also called residual stresses of the first kind, are homogeneous on a microscopic scale. Micro residual stresses are stresses which can vary strongly on a microscopic scale. Within one grain these stresses annihilate one another.

7.3.2 *Training procedures*

The processing to induce the TWME consists in general of thermomechanical cycling through the transformation region to acquire the cold, martensitic shape and is therefore referred to as training.[33,38,43,62,65,69–71,73–84] The training also results in concomitant effects, e.g. changes of the transformation temperatures and hysteresis, and residual deformations of the hot, parent shape.[33,43,65,69,82,84]

Next to the cycling procedures, the following one-time procedures have been reported. Remnants of preferentially oriented variants can be stabilized by holding a constrained or stressed sample at a temperature T_h, above the nominal A_f, for a sufficiently long time. The TWME obtained by this procedure is attributed to the growth of those remnants.[85–91] Ageing of a parent sample at sufficiently high temperatures and stresses can also result in reversible shape effects, such as the all-round shape memory effect.[92–97] The bainitic precipitates, formed during ageing, induce a residual stress field. The observed reversible shape effect is attributed to this residual stress field and also to the interaction between the martensite formation and the preferentially oriented precipitates. A TWME of small magnitude can be also obtained by a single, sufficiently high plastic deformation of the martensitic phase.[70–72,82] However, the disadvantages of these one-time procedures are numerous: large residual strains of the hot shape, large shifts of the transformation temperatures, strong dependence on the stabilization or ageing temperatures, stresses and times, etc.

Most training procedures are based on the repetition of transformation cycles from the parent phase to preferentially oriented martensite. Examples of these training procedures are: repetition of the one-way memory effect, temperature cycles at a constant strain or constrained cycling, temperature cycles at a constant stress, and superelastic cycling. Combinations and variants of these procedures can also be applied. As a result, 'new' procedures are regularly reported in the literature. In most of those publications, the new aspects of the procedure are emphasized, but almost no attention is paid to the points of similarity with previously described procedures. In addition, it is often stated that those 'new' training procedures are 'better' or 'more efficient' according to some rather arbitrary criterion. It is however clear that there can be only one criterion to determine the optimum training procedure: the optimum training procedure and the optimum parameters within this procedure have to result in optimum TWME behaviour, i.e. a combination of maximum magnitude, reproducibility and stability of the TWME and a minimum change of the hot shape and of the transformation temperatures.[84]

The mechanism of the repetitive training procedures can be easily explained from the thermodynamic description.[33,84] The training cycling results in the formation of complex dislocation arrays which have the lowest energy in the repeatedly induced 'trained' variants. Since the density of dislocation arrays increases to a saturation value during training cycling, the thermodynamic favouring of the trained variants also increases to a saturation value. It follows that during training cycling the magnitude of the TWME increases towards the strain induced during training. Analogously, the stability and the reproducibility of the TWME increase to a saturation value during training cycling.

7.4 Constrained recovery – generation of recovery stresses

7.4.1 Introduction

In the case where the recovery of the hot shape is impeded during heating, recovery stresses are generated. Four new parameters have to be introduced to describe this shape memory property (Fig. 7.1): the contact strain ε_c, the contact temperature T_c, the recovery stress σ_r and the stress rate $d\sigma_r/dT$. The generation of recovery stresses starts from a macroscopic deformation in the martensitic state. During subsequent heating, free recovery occurs till a temperature T_c. The recovery of the remaining deformation, characterized by the contact strain ε_c, is impeded by an external mechanical obstacle. So, from the temperature T_c recovery stresses are generated at a nearly constant stress rate $d\sigma_r/dT$, often described by a Clausius–Clapeyron equation. This shape memory property is the basis of some very successful shape memory applications, e.g. the well-known tube couplings.[98,99] Nevertheless, the number of publications on this shape memory effect is rather limited.[98–109] Although review publications can give the impression that the generation of recovery stresses is well understood, a careful analysis has shown that the scientific understanding of this property is still strictly limited,[33] which can be partly explained by the shortage of experimental results. Moreover, making an exception for the results of Duerig et al.[103,107] and Proft et al.,[98] the published experimental results are of doubtful quality. Indeed, in most studies a hard tensile machine has been used, in which case the thermal dilatations of the crosshead arms can have such a dominating effect that the measured results are unreliable.[65] For this and other reasons, a fundamental model, which allows one to predict quantitatively the recovery stress generation by means of a limited number of input parameters, is still missing.

7.4.2 A thermodynamic description of recovery stress generation

The total strain ε_t can always be subdivided into a transformation component ε_{tr}, an elastic component ε_e and a thermal dilatation component ε_{th}. During constrained recovery, the total strain ε_t is equal to the contact strain ε_c:

$$\varepsilon_c(T) = \varepsilon_t(T) = \varepsilon_{tr}(T) + \varepsilon_e(T) + \varepsilon_{th}(T). \tag{7.16}$$

The 'adapted' contact strain ε_{ct} is defined as:

$$\varepsilon_{ct} = \varepsilon_c - \varepsilon_{th}. \tag{7.17}$$

An ideal case is considered in which the mechanical obstacle is completely rigid and in which the pure thermal dilatation of the shape memory element is equal to the thermal dilatation of the obstacle. Under these conditions, ε_{ct} is independent of temperature and stress, i.e. ε_{ct} is constant during the generation of recovery stresses:

$$\varepsilon_{ct}(T) = \varepsilon_{tr}(T) + \varepsilon_e(T) = \text{constant.} \tag{7.18}$$

From Eq. (7.18) it follows that ε_e increases gradually and ε_{tr} decreases to the same extent during constrained heating. The resulting gradual decrease of the transformation fraction $[x]$ implies that the conditions to apply the Clausius–Clapeyron equation are not fulfilled. However, it can be shown that the Clausius–Clapeyron equation (Eq. (7.11)) is still a close approximation.[33]

If ε_{ct} is sufficiently high, plastic deformation occurs during constrained heating. The yield stress σ_y is in that case an upper limit to σ_r. If ε_{ct} is sufficiently small ($\varepsilon_{ct} < \sigma_y /E$), the sample can retransform completely during constrained heating. In that case a maximum recovery stress σ_{rm}, equal to the product $\{E^*\varepsilon_{ct}\}$, is obtained from a temperature T_m corresponding to complete retransformation.

The variant arrangement, and so also $\Delta\varepsilon_{tr\Sigma A}$ and $\{d\sigma_r/dT\}$ are determined to a large extent by the *thermomechanical history*. An important example in this respect is that training can have an important effect on $\{d\sigma_r/dT\}$. Also $\Delta\varepsilon_{tr\Sigma A}$ decreases in general with increasing *transformation fraction* x.[63,64] Consequently, $\{d\sigma_r/dT\}$ becomes smaller with higher T_c. Since the transformation fraction x decreases gradually during recovery stress generation, $\{d\sigma_r/dT\}$ becomes smaller with higher σ_r. This theoretical analysis is supported by the experimental results in Fig. 7.4.

Figure 7.5(a) shows the events which occur corresponding to the thermodynamic model when constrained heating is followed by constrained cooling. The generation of recovery stresses during constrained heating corresponds to a thermodynamic two-phase equilibrium *from martensite to the parent phase.*

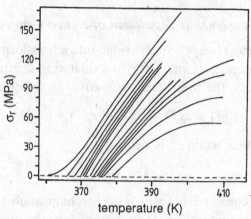

Fig. 7.4. Two-dimensional representation of 13 recovery stress experiments starting from different contact temperatures on a trained Cu–Zn–Al based sample. (After Stalmans[33])

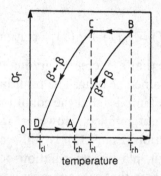

Fig. 7.5(a). Schematic representation of the increase and decrease of recovery stresses during constrained heating and cooling. Recovery stresses are generated during heating (A–B) in the temperature region $\{T_{ch}-T_{rh}\}$, corresponding to a thermodynamic equilibrium from martensite to the parent phase ($\beta' \rightarrow \beta$). Recovery stresses are decreased during cooling (C–D) in the temperature region $\{T_{rl}-T_{cl}\}$, corresponding to a thermodynamic equilibrium from the parent phase to martensite ($\beta \rightarrow \beta'$). Both temperature regions are separated by a hysteresis region, represented by $\{T_{rh}-T_{rl}\}$ and $(T_{cl}-T_{ch})$, in which σ_r is constant.

After reaching the temperature T_{rh}, the temperature is decreased at constant ε_{ct}. A temperature hysteresis has to be overcome before a thermodynamic two phase equilibrium *from the parent phase to martensite* is attained at a temperature T_{rl}. The recovery stress σ_r remains constant in the hysteresis region $\{T_{rl} - T_{rh}\}$. The two-phase equilibrium *from the parent phase to martensite* is preserved during further cooling and σ_r decreases to zero; it can be easily deduced that the relationship between σ_r and T_r is again described by Eq. (7.11). Also during constrained heating after constrained cooling, a hysteresis

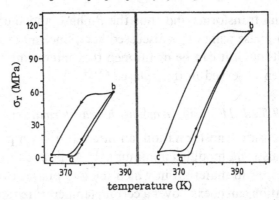

Fig. 7.5(b). Experimental results showing the recovery stress generation and decrease during thermal cycling at a constant contact strain ε_{ct}. The trained Cu–Zn–Al based sample is heated (a → b), followed by cooling (b → c) and heating again (c → b). (After Stalmans[33])

region $\{T_{ch} - T_{cl}\}$ has to be overcome. An experimental confirmation is shown in Fig. 7.5(b).

Recently, it has been shown that, in the case of *trained* samples, recovery stresses are also generated when an external constraint prevents the sample from returning to the cold shape on cooling.[33,38] These stresses are called 'cold shape' recovery stresses to contrast with the above mentioned 'hot shape' recovery stresses. This new property can be easily explained, similarly to the work output during cooling, by use of the thermodynamic description.

7.5 The high damping capacity of shape memory alloys

7.5.1 Introduction

The internal friction occurring in shape memory alloys during transformation and in the martensitic state has been studied in detail. It appears that Cu–Zn–Al alloys especially exhibit the highest damping capacity of all hidamets which makes them attractive for specific applications. Several reviews have already been dedicated to the particular damping behaviour of materials exhibiting a thermoelastic or non-thermoelastic martensitic transformation.[110-114] The most relevant part of the present state of knowledge, mainly related to Ni–Ti and Cu-based shape memory alloys, will be explained further.

7.5.2 The internal friction (IF) behaviour of shape memory alloys

Two main temperature regions have to be considered separately: the temperature region in which the material is completely martensitic and the one in which the material transforms and thus the β-phase and the martensite are coexisting. The β-phase will not be discussed here, since no high damping has been observed although it can be mentioned that interesting relaxation phenomena have been observed in this phase.[115-117]

7.5.2.1 IF during martensitic transformation

During the martensitic transformation, an internal friction peak is observed concurrent with a strong modulus minimum.[118] As pointed out by Bidaux *et al.*[119] and Stoiber,[120] in materials in which the two phases connected by the phase transformation can coexist over a certain temperature range, one should consider three separate contributions to the total internal friction, see Fig. 7.6.

$$Q_{tot}^{-1} = Q_{Tr}^{-1} + Q_{PT}^{-1} + Q_{int}^{-1}. \tag{7.19}$$

Q_{Tr}^{-1} is the transient part of Q_{tot}^{-1}, and it exists only during cooling or heating ($dT/dt \neq 0$). It depends on external parameters like temperature rate (dT/dt), resonance frequency (f) and oscillation amplitude (σ_0). Q_{Tr}^{-1} depends on the transformation kinetics and is therefore proportional to the volume fraction which is transformed per unit of time. Q_{PT}^{-1} is related to mechanisms of the phase transformation (PT), which are independent of the transformation rate, such as the movement of parent–martensite and martensite–martensite interfaces. Q_{PT}^{-1} exhibits a small peak when the interface mobility is highest. Q_{int}^{-1} is composed of the IF contributions of each phase and is strongly dependent on microstructural properties (interface density, vacancies), especially in the martensitic phase. The observation of the damping peak value during transformation is mainly due to Q_{Tr}^{-1}.

Generally the internal friction or damping capacity of shape memory alloys is measured at a constant heating or cooling rate and thus a significant damping peak appears during transformation. However at $T = 0$, thus at

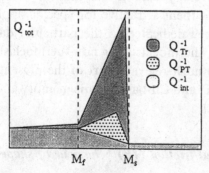

Fig. 7.6. Schematic representation of Q^{-1} and its contributions during martensitic phase transformations.

constant temperature, Q_{Tr}^{-1} becomes zero. This is also actually seen during the experiment and the remaining damping capacity remains a summation of Q_{int}^{-1} and Q_{PT}^{-1}. As a consequence, the damping capacity becomes a function of the volume fraction of the martensite and the very high damping capacity *during* transformation is lost. This means that for applications where a continuous vibration is applied at constant temperature a 100% martensite condition offers more stable damping.

In the case of impact loadings with very low frequency, the two-phase region can be interesting since the martensite will now be stress-induced concurrent with an exothermic heat effect.

7.5.2.2 *The internal friction in martensite*

So far, it is generally accepted that the reversible motion of the interfaces between the different variants is responsible for the observed behaviour of nonlinear damping.[111,112] The origin of the energy loss is not so clear. Electron microscopic observations revealed some specific contrasts at the interfaces which are explained as interface dislocations.[123] These dislocations form an ordered dislocation network. Moreover, the dislocations are probably more mobile in the martensite than in the β-phase which can also contribute to higher damping. From the latter model, it can be concluded that no high IF can be measured in a sample consisting of one monovariant, that is without interfaces. From the few publications treating the IF of a monovariant, the hypothesis could not be proved exactly. Morin *et al.*[121] who measured the IF with a low frequency pendulum conclude indeed that the interfaces are a necessary condition for high damping. Dejonghe *et al.*[122] however, measuring the IF by a cantilever, detected first a high damping in the martensite but this damping diminished with thermal cycling. So these authors conclude that the high damping is also related to other defects, specific for the martensitic structure, but the mobility of these defects (dislocations, stacking faults) can be decreased by pinning effects. In order to obtain a unique conclusion, it would be necessary to repeat the experiments of Morin and Dejonghe with monovariants all having a different orientation.

7.5.2.3 *How large is the damping capacity?*

First of all, the external parameters temperature, time, frequency, but most importantly the amplitude can change the damping capacity. The type of material, grain size, martensite interface density and defect structure are important internal variables.[124]

In any case, one may state that the martensite of Cu-based alloys and Ni–Ti show a damping capacity of at least an order of magnitude higher than classic materials such as steels. For high amplitude ($\varepsilon \approx 10^{-4}$), the loss factor in martensite can be of the order of 6 to 8%. During impact loading 10% and more can be obtained. This loss factor decreases to about 2 to 4% for amplitudes in the order of 10^{-5}.

7.6 Cycling effects, fatigue and degradation of shape memory alloys

7.6.1 Introduction

The specific functional properties of shape memory alloys necessitate an extension of the usual definition of fatigue.

Three different types of fatigue have to be considered in the case of shape memory alloys.

(i) Failure by fracture due to stress or strain cycling at constant temperature.[125,142]
(ii) Changes in physical, mechanical and functional properties such as the transformation temperatures, the transformation hysteresis, two-way memory effect, ... due to pure thermal cycling through the transformation region.[53,54,59,143–147]
(iii) The degradation of the shape memory effect due to stress, strain or temperature cycling in or through the transformation region.[148–155]

The three main parameters to be considered in the study of the global lifetime of a shape memory alloy are thus temperature, stress and the macroscopic shape strain during deformation. It might be important for the latter to distinguish between the transformation induced plasticity, that is, the shape strain under load during the forward transformation, and the spontaneous shape strain due to the one- or two-way memory effect.

The origin of the global lifetime or fatigue behaviour of shape memory alloys is due to accumulation of defects and structural changes like the change in order of the β-phase or martensite or the formation of other types of martensite. All types of changes will have an influence on the transformation temperatures, the transformation hysteresis, the reproducibility of the one- and two-way memory effects, and the number of cycles before fracture.

The reliability of a shape memory device depends on its global lifetime performance. Time, temperature, stress, transformation strain and the number of transformation cycles are the important controlling external parameters. Important internal parameters that determine the physical and mechanical properties are mainly: the alloy system, the alloy composition, the type of transformation and the lattice structure including defects. Those parameters are controlled by the thermomechanical history of the alloy. Changes of those internal parameters give rise to effects such as increased frictional forces, stabilization of martensite, defect generation or annihilation, and precipitation.

7.6.2 *Failure by fracture*

Classic fatigue implies accumulation of defects, formation of cracks and crack propagation until final rupture. This type of behaviour is represented by the well-known Wöhler curves, indicating the stress at which the material will fail as a function of the number of cycles. Figures 7.7[139] and 7.8[140] illustrate this behaviour for respectively Ni–Ti and Cu–Zn–Al. The big scatter in the Wöhler curves provides, in fact, poorly reliable data and contains mainly qualitative information.[140,142] For example fine grained alloys have a higher fatigue resistance than coarse grained alloys. Other reasons for low fatigue life might be the presence of inclusions, segregation of impurities to the grain boundaries or the quality of the surface. Other factors that make a comparison between the present results difficult are the differences in experimental conditions such as the R-value (the ratio of the minimum (S_{min}) and maximum stress (S_{max}) of the applied stress cycle) or the use of stress or strain controlled fatigue tests.

Other information becomes available using (da/dN, $\log\Delta K$) plots. da/dN is the change in crack length a versus the number of cycles N. ΔK is the stress intensity factor. $\Delta K = K_{max} - K_{min}$ in which $K_{max} = C \cdot S_{max} \cdot \sqrt{\pi a}$ and $K_{min} = C \cdot S_{min} \cdot \sqrt{\pi a}$.

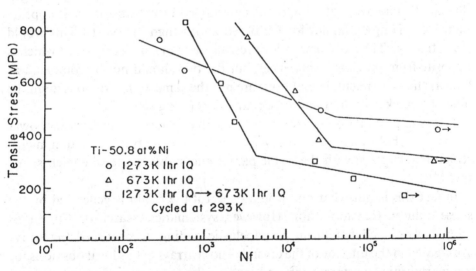

Fig. 7.7. Effect of heat-treatment on the fatigue life of the Ti–50.8 at% Ni alloy, tested at room temperature. (After Miyazaki and Otsuka[139])

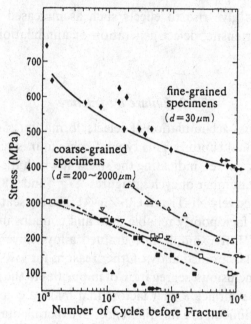

Fig. 7.8. Fatigue life in fine- and coarse-grained Cu–Zn–Al specimens. (After Janssen et al.[140])

Although this type of presentation is becoming an increasing standard for fatigue design, only a few results have been reported in the case of shape memory alloys[126,139,141] and an example is illustrated in Fig. 7.9.[139]

An important value derived from this type of curve is the lower fatigue threshold value ΔK_{TH} at $da/dN = 0$, determined by extrapolation. It appears that ΔK_{TH} is much higher for Cu-based alloys than for Ni–Ti. This would mean that Ni–Ti has a worse fatigue resistance than Cu-based alloys, which is opposite from previous experiences, but the two should not be compared as long as the experimental conditions are not the same. Other reasons might be different crack initiation and crack growth mechanisms.[136,142]

A very important observation however is that faster growth rates and lower fatigue threshold values are observed in the two-phase transformation micro-structure compared with the stable parent and stable martensite microstructure.[141]

In fact this is not what one would expect if the strain is generated by the stress induced transformation. However, systematic research on Cu–Zn–Al alloys has revealed that during pseudoelastic loading, dislocation defect arrays are created in the interior of the crystal. Those arrays act as local obstacles for the martensitic transformation and lead to the formation of extrusions and holes at the surface, which, with increasing number of cycles, join to form

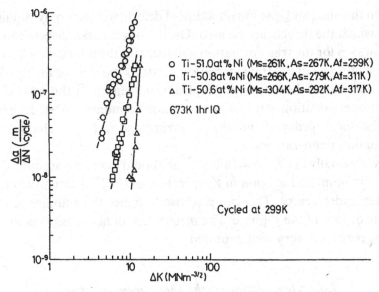

Fig. 7.9. Crack propagation rate at 299 K as a function of range of stress intensity factor for various Ti–Ni alloys. (After Miyazaki et al.[139])

continuous cracks of about 1 μm width.[125] The dislocation arrays initially change drastically the hardening slope of the stress induced transformation.[40,125] In Cu–Al–Ni single crystals, one author observed that the cracks initially grew parallel to the stress induced martensite plates. Other authors concluded that crack nuclei occur at interphase boundaries[131] or martensite–martensite interfaces.[134] In polycrystals, multiple nucleation cracks occur at grain boundaries which is similar to what happens in regular metals at high plastic strain.[136]

7.6.3 Thermal cycling

Thermal cycling through the martensitic transformation induces defects. Extensive literature surveys, especially for Cu-based alloys, can be found in Refs. [48,144,147,157].

Concerning the changes in the macroscopic parameters of the martensitic transformations, such as the transformation temperatures or the transformation hysteresis, the situation is less clear. Some authors report an increase of M_s[53,126] while others report a decrease, even for the same type of alloy.[147] Similar contradictions exist for the transformation temperatures A_s, A_f and M_f as for the hysteresis, which decreases[53,126] or increases.[59,156]

Accepting that the experimental observations are correct, one has to conclude that several mechanisms, active during thermal cycling, have opposite effects. On the one hand, the introduction of defects will generally stabilize the phase in which the defects are created. On the other hand, defects can act as nucleation sites for the transformation and increase the internal friction forces so that the progress of the transformation becomes more difficult. The density and configuration of the defects might also be influenced by the grain size, due to high accommodation stresses at the grain boundaries. Defects can also change the (local) order of the phases, giving rise to (local) changes of the transformation temperatures.

Finally, especially in Cu-based alloys, one should take into account that the absolute minimum and maximum temperature of the cycle might control the equilibrium order degree. This can give rise to either the stabilization of the martensite or that of the β-phase. The mechanism of martensite stabilization, especially, is not yet very well explored.

7.6.4 Degradation of the shape memory effect

It has been found that the global degradation behaviour is influenced by a complex combination of internal and external parameters. Internal parameters are: the alloy system (Cu-based alloys are more prone to degradation than Ni–Ti alloys); the alloy composition; the type of transformation (i.e. martensite *vs* R-phase); the lattice structure, including defects. External parameters are: the thermomechanical treatment; the training procedure; the applied stress; the imposed shape memory strain; the amplitude of temperature cycling; the average absolute temperature.

To identify the different mechanisms of degradation, a closer look at the dimensional changes of both the cold and the hot shape is required. Stalmans et al.[65] analysed this in detail and several parameters were identified. In the course of training and degradation, three different types of strain are important: TWME, or the difference in strain between the hot and the cold shape; TWMS (Two-Way Memory Strain), or the difference between the strain of the martensite after *n* cycles and the strain before cycling; finally RD, the Remained Deformation of the β-phase after *n* cycles. Figure 7.10 shows the evolution of RD, TWMS and TWME for a Cu–Zn–Al alloy. Similar results are presented by Rodriguez and Guénin.[158]

From Fig. 7.10 it is clear that the initial degradation of the two-way memory effect is due to an 'apparent' permanent strain in the β-phase. Indeed, the total permanent strain RD is partially due to real plastic deformation of the lattice and partially due to stabilization of the martensite.[159] The latter part can be restored by heating to a temperature higher than the maximum temperature used during the (training) cycles.

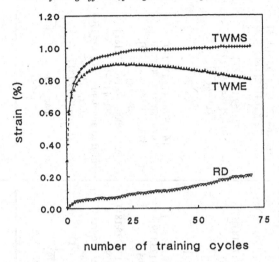

Fig. 7.10. The evolution of the TWME (Two-Way Memory Effect), the TWMS (Two-Way Martensitic Strain) and the residual deformation of the parent shape RD.

If cycling is continued to hundreds or thousands of cycles, the cold shape can also decline. In other words, the TWMS decreases. Several mechanisms can contribute to the degradation. For Cu–Zn–Al, Friend and Miodownik[150] observed locally α' martensite, stress induced during cycling. This increasingly disrupts the martensite morphology so that the strain output degrades. A second part might be related to the irreversible dislocation debris generated during thermal cycling. The defect accumulation may also account for the increasing strain of the hot shape. It may also destroy the initial nuclei that normally trigger the appropriate variants. This might be explained by the increase of the amount of misoriented variants.[160]

Some observations revealed that the rate of degradation is also dependent on the amount of the initial two-way memory effect.[161] A high TWME will degrade relatively faster than a small TWME.[187] Other influences are the method of training,[155] the addition of alloying elements,[146] or thermal treatment[152] as explored in Ni–Ti alloys.

An important remark is given by Suzuki.[161] He noticed that the degradation in Ni–Ti alloys is related to the size of the hysteresis, the smaller the hysteresis, the lower the degradation. In this respect, the R-phase transition, which shows a very small hysteresis would be the best choice if the number of cycles has to approach 10^6. When the cooling is stopped after the R-transformation but before the martensite starts to grow, an improved lifetime is registered.[161]

Table 7.1. *Property values*

property	unit	NiTi	CuZnAl	CuAlNi
physical				
melting point	°C	1240–1310[169,170,172,173] 1250[179]1300[185]	950–1020[169,170,173,181] 1020[179]	1000–1050[169,170,173,181] 1050[179]
density	10^3 kg m^{-3}	6.4–6.5 [167,169,170,172,173,175,183] 6.45[179]6.52[182] 6.5[184–186]	7.8–8.0[167,169,170,173,175,181,183] 7.64[164]7.9[179] 8.0[184]7.5[186]	7.1–7.2 [167,169,170,173,175,181,183] 7.12[164]7.2[184] 7.15[179]
thermal conductivity 20°C	W m^{-1} K^{-1}			
parent		10–18[169,172,173,179,180] 18[183]	120[164,169,173,183] 84[181]	75[169,173,183] 30–43[164]69[181]
martensite		18[170,185] 8.6[170,185]	120[170,170]	75[170,179]
coeff. of thermal expansion	10^{-6} K^{-1}	6.6–10[179]	17[179]	17[179]
parent		11[170,185]10[170,173,182]	16–18[169,170,173,179]	16–18[169,170,173,179]
martensite		6.6[169,170,173,182,185]		
specific heat	J kg^{-1} K^{-1}	470–620[169,170,173] 490[179]450[172]	390[169,170,173,179] 400[164]	400–480[169,170,173] 373–574[164]440[179] 480[181]
transformation enthalpy	J kg^{-1}	3200[169,170,173] 19 000[172]28 000[179]	7000–9000[169,170,173,181] 7000[179]	7000–9000[169,170,173,181] 9000[179]
corrosion performance		similar to 300 series stainless steel [170,185]	similar to aluminium bronzes [170] poor[183]	similar to aluminium bronzes [170] fair[183]
biological compatibility		comparable to pure Ti[172] excellent[179,183]	excellent[181] fair[179]	good[179,181]
wear resistance		excellent[179] + +[183] good	bad[179] –[183]	bad[179] –[183]
electromagnetic				
resistivity	10^{-6} Ωm			
parent		0.5–1.1[169,170,173,179,183] 0.7–1.0[167,175]0.8[182]	0.07–0.12[169,173,179,181,183] 0.85–0.97[164]0.08–0.13[167,175]	0.10–0.14[169,173,179,181,183] 0.11–0.13[164]0.11–0.14[167,175]
martensite		1.0[170,185] 0.8[170,185]	0.07[170] 0.12[170]	0.1[170] 0.14[170]
thermo-electric power	10^{-6} V K^{-1}	9–13(mart)[169,173] 5–8(aust)[169,173]		
magnetic permeability		< 1.002[170,172,185]		
magnetic susceptibility	emu g^{-1}	3.10^6[170,185]		

mechanical Young's modulus parent	GPa	70–98[170]95[179,183]	70–100[170,172,179,181]80[183]	80–100[170,179,181]100[183]
martensite		70[172]98[169,173] 97[182,185]	70–100[169,173]	80–100[169,173]
G (parent)	GPa	27[172]	72[164]	80[164]
yield strength parent	MPa	410[185] 200–800[169,170,173] 100–600[172]410[172] 150–300[169,170,173] 50–300[172]	70[164] 350[164]150–300[170]	150–300[170]
martensite			80[164] 150–300[169,173]	150–300[169,173]
ultimate tensile strength	MPa	800–1000[167,175,179,183] 900–1500[184]860[185] 800[186]	400–700[167,175,184]600[164] 800–900[179]700–800[181,183] 500[186]	700–800[167,175,184] 500–800[164]1000–1200[181,183] 1000[179]
martensite		800–1100[169,173] 700–1100(annealed)[170,172] 1300–2000(not anneal.)[170,172] 860[182]	700–800[169,170,173]	1000–1200[169,170,173]
elongation at failure	%	40–50[167,175]50[183]	10[173]10–15[167,175,181,184] 15[183]	5–6[167,175]8–10[181] 12[183]4–6[184]
parent		30–50[184] 15–20[182,185]		
martensite		40–50[169,170,173] 20–60[172]30–50[179]	10–15[169,170,173] 15[179]	8–10[169,170,173,179]
fatigue strength $N = 10^6$	MPa	350[169,170,173,179]	270[169,170,173,179,181] 50–100[169,170,173,181]	350[169,170,173,179,181] 25–60[169,170,173,181]
grain size	μm	1–10[170]50–100[169,170,173] 20–100[179]	50–150[179]	30–100[179]
elastic anisotropy $2 C_{44}(C_{11}-C_{22})^{-1}$		2[170]	15[164,170]	13[170]12[164]
shape memory transformation temperatures[a]	°C	– 100 to 100[172]120[167,175,184] – 100 to 120[169,173] – 40 to 100[173] – 100 to 110[179] – 200 to 100[170,185] – 20 to 80[183] 200 to 300 (NiTiPd)[173] 580(NiTi50at%Pd)[163] 108 to 170 (NiTiZr)[177]	– 200 to 120[169,170,173]150[167] 120[175,184] – 190 to 100[176] – 200 to 110[179] – 170 to 110[181] – 200 to 100[183]	– 200 to 170[169,170,173] 200[167]170[175,184] – 140 to 200[176] – 150 to 200[179,181] 50 to 180[183] – 50 to 180 (CuAlNiMnTi)[173]

Table 7.1. (*cont.*)

property	unit	NiTi	CuZnAl	CuAlNi
hysteresis	°C	20–30[170,172,175] 30[169,173,179,183] 4 (NiTi20%Cu)[162] 66(NiTi9at%Nb)[178] >120(NiTi9at%Nb)[174]	5–20[173]10–20[169,170,173,181] 10–25[164]7–15[175] 15[179,183]	20–30[169,170,173,181] 20–40[175] 20[179]15[183]
one-way memory strain	%	8[167,169,170,173,175,182, 183,184,185]7[179]6[186]	6[173]5[167,169,170,173,181, 183] 4[175,179,184,186]	5[167,175,183,184] 6[169,170,173,179,181] 8(single crystal)[180]
$N < 10^2$		6–8[172]	4[164]	4[164]
$N < 10^5$		2[172]		
$N < 10^7$		0.5[172]		
two-way memory strain	%	5[167,175]3.2[179]4[186]	4[173]2[164,167,175,186]0.8[179]	2[164,167,175,186][179] 1.2[169,170,173,181]1[183]
$N < 10^2$		6[169,170,173,183]	1[169,170,173,181,183]	0.8[169,170,173,181,183]
$N < 10^5$		2[169,170,173,183]	0.8[169,170,173,181,183]	0.5[169,170,173,181,183]
$N < 10^7$		0.5[169,170,173,183]	0.5[169,170,173,181,183]	
maximum temperature	°C	400[167,169,175,170,173,186,179]	160[167,175]160–200[169,170,173,181] 200[173,186]150[179]	300[167,169,170,173,175, 179,181]
damping capacity[b] (1 hr)	%SDC	15[169,170,173]20[179]	30[169,170,173]85[179]	10[169,170,173]20[179]
pseudoelastic strain[c] single crystal	%	10[169,170,173]	10[169,170,173,181]	10[169,170,173,181]15[180]
polycrystal		4[169,170,173]	2[169,170,173,181]	2[169,170,173,181]
superelastic energy storage	J/g	8[173]10[168] 6.5[168]	5[168] 1.8[168]	1[168]
max. recovery stress	MPa	600–800[170,172] 600[173,183]500–900[198] 700(NiTi9at%Nb)[165]	700[173]500[183] 550–650(CuZnAlMn)[98] 400(CuZn10wt%Al5wt%Mn)[165]	600[183] 300–400[98]

recovery strain	%	8[98] 8(NiTi9at%Nb)[165] 12[171]3–20[168]4–20[98] 3.5(NiTi9at%Nb)[165]	3.5(CuZnAlMn)[98] 3.5(CuZn10wt%Al5wt%Mn)[165] 2.5[171]2.5–6[33]2–5[98] 2(CuZn10wt%Al5wt%Mn)[165]	2.0[98]
stress rate	MPa K^{-1}	4[166]1[172]	1[166]	
work output	J/g			
economic				
melting, casting and composition control		difficult, in vacuum [169,170,183]	fair[169]air[183] easier[170]	fair[169]air[183] easier[170]
forming (rolling, extrusion)		hot, difficult[169,170] very difficult[183]	warm, easy[169,170] fairly easy[183]	hot, easy[169] hot, difficult[170,183]
cold-working		fair[169] difficult[170]	restricted [169,170]	very difficult[169] not possible[170]
machinability		difficult [169,170,183] bad[172]	very good[169,170] easy[183]	good[169,170] rather difficult[183]
cost ratios[d]		10[169] 100[170]	1[169] 1.0–10[170]	2[169] 1.5–20[170]

[a] in most cases A_s
[b] or short time
[c] dependent on frequency and amplitude
[d] varies greatly with shape, required quantities, etc.

7.7 Property values

Table 7.1 summarizes data on physical, mechanical, economic and functional properties of the three main classes of shape memory alloys: NiTi, CuZnAl and CuAlNi. These data were collected from the literature but should be handled critically. Indeed, the reliability of the data is sometimes rather poor and often no information on how the data were collected is available. Some values much depend on the reference itself and a few times are even contradictory. Therefore the authors would like to promote standard acceptance tests to define uniquely the different types of properties. Nevertheless, the authors hope that this table can be a handsome reference to be used by critical persons.

Acknowledgements

J. Van Humbeeck (Research director F.W.O.) and R. Stalmans (Postdoctoral Researcher F.W.O.) acknowledge the F.W.O. (National Fund for Scientific Research, Belgium).

References

1. M. M. Reyhani and P. G. McCormick, *Scr. Metall.*, **20** (1986) 571.
2. K. Takezawa and S. Sato, *Trans. JIM*, **29** (1988) 894.
3. D. Favier, Doctorate Thesis, L'Institut National Polytechnique de Grenoble, Grenoble (1988).
4. I. Müller, in *The Martensitic Transformation in Science and Technology*, ed. E. Hornbogen and E. Jost, (DGM Informationsgesellschaft, Oberursel, 1989) p. 69.
5. L. Lü, E. Aernoudt, P. Wollants, J. Van Humbeeck and L. Delaey, *Z. Metallkd.*, **81** (1990) 613.
6. K. Tanaka, *Res. Mechanica.*, **18** (1986) 251.
7. J. R. Patel and M. Cohen, *Acta Metall.*, **1** (1953) 531.
8. H. Warlimont, L. Delaey, R. V. Krishnan and H. Tas, *J. Mater. Sci.*, **9** (1974) 1545.
9. P. Wollants, M. De Bonte and J. R. Roos, *Z. Metallkd.*, **70** (1979) 113.
10. P. Wollants, M. De Bonte, L. Delaey and J. R. Roos, *Z. Metallkd.*, **79** (1979) 146.
11. N. S. Kosenko, A. L. Roytburd and L. G. Khandros, *Phys. Met. Metall.*, **44** (1979) 48.
12. F. Falk, *Acta Metall.*, **28** (1980) 1773.
13. P. Wollants, M. De Bonte and J. R. Roos, *Z. Metallkd.*, **74** (1983) 127.
14. P. Wollants, Doctorate Thesis, Catholic University of Leuven, Department of Metallurgy and Materials Science, Heverlee (1983).
15. M. Kato and H. R. Pak, *Phys. Stat. Sol. B*, **123** (1984) 415.
16. M. Kato and H. R. Pak, *Phys. Stat. Sol. B*, **130** (1985) 421.
17. J. S. Cory and J. L. McNicholls, *J. Appl. Phys.*, **58** (1985) 3282.
18. J. L. McNichols and J. S. Cory, *J. Appl. Phys.*, **61** (1987) 972.
19. J. Ortin, *Thermochimica Acta*, **121** (1987) 397.
20. J. Ortin and A. Planes, *Acta Metall.*, **36** (1988) 1873.
21. A. Planes, J. L. Macqueron and J. Ortin, *Phil. Mag. Letters*, **57** (1988) 291.
22. J. Ortin and A. Planes, *Acta Metall.*, **37** (1989) 1433.
23. J. Ortin, in *The Science and Technology of Shape Memory Alloys*, ed. V. Torra (Impresrapit, Barcelona, 1989) p. 142.

24. J. Ortìn and A. Planes, in *The Martensitic Transformation in Science and Technology*, ed. E. Hornbogen and E. Jost (DGM Informationsgesellschaft, Oberursel, 1989) p. 75.
25. J. Ortìn and A. Planes, *J. de Phys. Iv, Colloque C4*, addendum to *Journal de Physique III*, **1** (1991) 13.
26. P. Wollants, J. R. Roos and K. Otsuka, *Z. Metallkd.*, **82** (1991) 182.
27. J. Ortìn, in *Proc. Int. Conf. on Mart. Transf. '93*, ed. C. M. Wayman and J. Perkins (Monterey Institute of Advanced Studies, Carmel, 1993) p. 305.
28. H. Tas, Doctorate Thesis, Catholic University of Leuven, Department of Metallurgy and Materials Science (1971).
29. J. De Vos, Doctorate Thesis, Catholic University of Leuven, Department of Metallurgy and Materials Science (1978).
30. M. Ahlers and H. Pops, *Trans. AIME*, **242** (1968) 1267.
31. W. Arneodo and M. Ahlers, *Acta Metall.*, **22** (1974) 1475.
32. H. Tas, L. Delaey and A. Deruyttere, *J. Less-Common Metals*, **28** (1972) 141.
33. R. Stalmans, Doctorate Thesis, Catholic University of Leuven, Department of Metallurgy and Materials Science, Heverlee (1993).
34. T. Saburi, S. Nenno, J. Hasunuma and H. Takii, in *New Aspects of Martensitic Transformation*, addendum to *Trans. JIM*, **17** (1976) 251.
35. T. A. Schroeder, I. Cornelis and C. M. Wayman, *Metall. Trans. A*, **7** (1977) 535.
36. K. Otsuka, C. M. Wayman, K. Naikai, H. Shimizu and K. Sakamoto, *Acta Metall.*, **24** (1976) 207.
37. J. Van Humbeeck, L. Delaey and A. Deruyttere, *Z. Metallkd.*, **69** (1979) 575.
38. R. Stalmans, J. Van Humbeeck and L. Delaey, *Mater. Trans. JIM*, **33** (1992) 289.
39. R. Stalmans, J. Van Humbeeck and L. Delaey, *Acta Metall. Mater.*, **40** (1992) 2921.
40. A. Ritter, N. Y. C. Yang, D. P. Pope and C. Laird, *Metall. Trans. A*, **10** (1979) 667.
41. C. Mai, G. Guénin, M. Morin and P. F. Gobin, *Mater. Sci. Eng.*, **45** (1980) 217.
42. D. Rios-Jara, M. Morin, C. Esnouf and G. Guénin, in Int. Conf. on Mart. Transf. '82, *J. de Phys.*, *Colloque C-4*, addendum to n°12 (1982) 735.
43. J. Perkins and R. O. Sponholz, *Met. Trans. A*, **15** (1984) 313.
44. D. Rios-Jara, M. Morin, C. Esnouf and G. Guénin, *Scr. Metall.*, **19** (1985) 441.
45. M. S. Andrade, J. Janssen and L. Delaey, *Metallography*, **18** (1985) 107.
46. M. Sade, R. Rapacioli and M. Ahlers, *Acta Metall.*, **33** (1985) 487.
47. M. Sade and M. Ahlers, *Scr. Metall.*, **19** (1985) 425.
48. D. Rios-Jara and G. Guénin, *Acta Metall.*, **35** (1987) 109.
49. M. Sade, A. Uribarri and F. Lovey, *Phil. Mag. A*, **55** (1987) 445.
50. M. Sade, F. Lovey and M. Ahlers, *Materials Science Forum*, **56–58** (1987) 934.
51. S. Kajiwara, in *New Aspects of Martensitic Transformation*, addendum to *Trans JIM*, **17** (1976) 81.
52. S. Kajiwara and T. Kikuchi, *Acta Metall.*, **30** (1982) 589.
53. J. Perkins and W. E. Muesing, *Met. Trans. A*, **14** (1983) 33.
54. Y. Nakata, T. Tadaki and K. Shimizu, *Trans. JIM*, **26** (1985) 646.
55. S. Miyazaki, Y. Igo and K. Otsuka, *Acta Metall.*, **34** (1986) 2045.
56. K. Shimizu, in *Shape Memory Alloy '86*, ed. Chu Yougi, T. Y. Hsu, T. Ko (China Academic Publishers, Beijing, 1987) p. 15.
57. K. Marukawa and S. Kajiwara, *Phil. Mag. A*, **55** (1987) 85.
58. T. Tadaki, M. Takamori and K. Shimizu, in *Shape Memory Alloy '86*, ed. Chu Yougi, T. Y. Hsu, T. Ko (China Academic Publishers, Beijing, 1987) p. 303.
59. T. Tadaki, M. Takamori and K. Shimizu, *Trans. JIM*, **28** (1987) 120.
60. T. Tadaki, Y. Nakata, K. Shimizu, *Trans. JIM*, **28** (1987) 883.
61. J. Pons, F. C. Lovey and E. Cesari, *Acta Metall. Mater.*, **38** (1990) 2733.
62. L. Delaey and J. Thienel, in *Shape Memory Effects in Alloys*, ed. J. Perkins (Plenum Press, New York, 1975) p. 341.
63. L. Delaey, G. Hummel and J. Thienel, *Encyclopaedia Cinematographica*, Film n°.E-2251 (1976).
64. C. M. Friend, *Scr. Metall.*, **20** (1986) 995.

65. R. Stalmans, J. Van Humbeeck and L. Delaey, *Acta Metall. Mater.*, **40** (1992) 501.
66. G. B. Olson and M. Cohen, *Scr. Metall.*, **9** (1975) 1247.
67. G. B. Olson and M. Cohen, *Scr. Metall.*, **11** (1977) 345.
68. L. Delaey and E. Aernoudt, in *Proc. Int. Conf. on Mart. Transf. '86*, ed. I. Tamura (The Japan Institute of Metals, Aoba Aramaki, 1987) p. 926.
69. G. Guénin, in *The Martensitic Transformation in Science and Technology*, ed. E. Hornbogen and E. Jost (DGM Informationsgesellschaft, Oberursel, 1989) p. 39.
70. A. Nagasawa, K. Enami, Y. Ishino, Y. Abe and J. Nenno, *Scr. Metall.*, **8** (1974) 1055.
71. J. Perkins, *Scr. Metall.*, **8** (1974) 1469.
72. J. Perkins and D. Hodgson, in *Engineering Aspects of Shape Memory Alloys*, ed. T. W. Duerig, K. N. Melton, D. Stöckel and C. M. Wayman (Butterworth–Heinemann, London, 1990) p. 195.
73. T. Saburi and S. Nenno, *Scr. Metall.*, **8** (1974) 1363.
74. R. J. Waseilewski, *Scr. Metall.*, **9** (1975) 417.
75. T. A. Schroeder and C. M. Wayman, *Scr. Metall.*, **11** (1977) 225.
76. J. Perkins, *Mater. Sci. Eng.*, **51** (1981) 181.
77. J. Perkins, in *Shape Memory Alloy '86*, ed. Chu Yougi, T. Y. Hsu, T. Ko (China Academic Publishers, Beijing, 1987) p. 201.
78. M. Sade, A. Hazarabedian, A. Uribarri and F. Lovey, in *Proc. Int. Conf. on Solid Phase Transformations*, ed. G. W. Lorrimer (Inst. of Metals, Cambridge, 1988) p. 279.
79. T. Todoroki, H. Tamura and Y. Suzuki, in *Proc. Int. Conf. on Mart. Transf. '86*, ed. I. Tamura (Japan Institute of Metals, Aoba Aramaki, 1987) p. 748.
80. M. M. Reyhani and P. G. McCormick, in *Proc. Int. Conf. on Mart. Transf. '86*, ed. I. Tamura (Japan Institute of Metals, Aoba Aramaki, 1987) p. 896.
81. P. Rodriguez and G. Guénin, in *The Martensitic Transformation in Science and Technology*, ed. E. Hornbogen and E. Jost (DGM Informationsgesellschaft, Oberursel, 1989) p. 149.
82. L. Contardo and G. Guénin, *Acta Metall. Mater.*, **38** (1990) 1267.
83. Y. Li and P. G. McCormick, *Acta Metall. Mater.*, **38** (1990) 1321.
84. R. Stalmans, J. Van Humbeeck and L. Delaey, in Proc. Int. Conf. on Mart. Trans. '92, ed. C. M. Wayman and J. Perkins (Monterey Institute of Advanced Studies, Carmel, 1993) p. 1065.
85. K. Takezawa and S. Sato, in *New Aspects of Martensitic Transformation*, addendum to *Trans JIM*, **17** (1976) 233.
86. K. Takezawa, K. Adachi and S. Sato, *Journal of the Japan Institute of Metals*, **43** (1979) 229.
87. K. Takezawa, H. Sato, Y. Abe and S. Sato, *Journal of the Japan Institute of Metals*, **43** (1979) 235.
88. R. Rapacioli, V. Torra, E. Cesari, J. M. Guilemany and J. R. Migel, *Scr. Metall.*, **22** (1988) 261.
89. B. G. Mellor, J. M. Guilemany, J. R. Migel, J. Fernandez, A. Amengual, F. Lovey and V. Torra, *Scr. Metall. Mater.*, **24** (1990) 241.
90. J. M. Guilemany, J. Fernandez and B. G. Mellor, *Scr. Metall. Mater.*, **24** (1990) 1941.
91. B. G. Mellor, J. M. Guilemany and J. Fernandez, *J. de Phys. IV*, *Colloque C4*, addendum to *J. de Phys. III*, **1** (1991) 457.
92. T. W. Duerig, J. Albrecht and G. H. Gessinger, *J. Metals*, **34** (1982) 14.
93. M. Nishida and T. Honma, *Scr. Metall.*, **18** (1984) 1293.
94. T. Honma, in *Proc. Int. Conf. on Mart. Trans. '86*, ed. I. Tamura (The Japan Institute of Metals, Aoba Aramaki, 1987) p. 709.
95. R. Kainuma and M. Matsumoto, *Scr. Metall.*, **22** (1988) 475.
96. S. S. Leu and C. T. Hu, *Scr. Metall.*, **23** (1989) 1925.
97. S. Edo, in Proc. Int. Conf. on Mart. Transf. '92 (Monterey Institute for Advanced Studies, Carmel, 1993) p. 965.
98. J. L. Proft and T. W. Duerig, in *Engineering Aspects of Shape Memory Alloys*, ed. T. W. Duerig, K. N. Melton, D. Stöckel and C. M. Wayman (Butterworth–Heinemann Ltd,

London, 1990) p. 115.

99. M. Kapgan and K. N. Melton, in *Engineering Aspects of Shape Memory Alloys*, ed. T. W. Duerig, K. N. Melton, D. Stöckel and C. M. Wayman (Butterworth–Heinemann, London, 1990) p. 137.

100. G. R. Edwards, J. Perkins and J. M. Johnson, *Scr. Metall*, **9** (1975) 1167.

101. J. Perkins, G. R. Edwards, C. R. Such, S. M. Johnson and R. R. Allen, in *Shape Memory Effect in Alloys*, ed. J. Perkins (Plenum Press, New York, 1975) p. 273.

102. H. A. Mohamed, *J. Mater. Sci. Letters*, **13** (1978) 2728.

103. T. W. Duerig and K. N. Melton, in *Shape Memory Alloy '86*, ed. Chu Yougi, T. Y. Hsu, T. Ko (China Academic Publishers, Beijing, 1987) p. 397.

104. K. Madangopal, S. Banerjee and M. K. Asundi, in *Shape Memory Alloy '86*, ed. Chu Yougi, T. Y. Hsu, T. Ko (China Academic Publishers, Beijing, 1987) p. 181.

105. K. Madangopal, R. Ganesh Krishnan and S. Banerjee, *Scr. Metall.*, **22** (1988) 1593.

106. Y. Furuya, H. Shimada, Y. Tanahashi, M. Matsumoto and T. Honma, *Scr. Metall.*, **22** (1988) 751.

107. T. W. Duerig and K. N. Melton, in *MRS Int'l. Mtg. on Adv. Mats.*, Vol. 9, (Materials Research Society, 1989) p. 581.

108. E. Cydzik, in *Engineering Aspects of Shape Memory Alloys*, ed. T. W. Duerig, K. N. Melton, D. Stöckel and C. M. Wayman (Butterworth–Heinemann, London, 1990) p. 149.

109. T. Borden, in *Engineering Aspects of Shape Memory Alloys*, ed. T. W. Duerig, K. N. Melton, D. Stöckel and C. M. Wayman, (Butterworth–Heinemann, London, 1990) p. 158.

110. R. De Batist, *J. de Phys.*, **C9**, Suppl. n. 12, 44 (1983), 39.

111. J. Van Humbeeck, *Proc. of Int. Symp. on Role of Interfaces on Material Damping*, ed. B. B. Rath and M. S. Misra, (ASM, 1985), p. 5.

112. J. Van Humbeeck, *Proc. of the Summer School on Internal Friction in Solids*, ed. S. Garczyca, L. B. Magalas (Wydawnictwo AGH, Krakow, 1984) p. 131.

113. J. Van Humbeeck, *Proc. of ICIFAS-9*, Beijing, June 17–20, 1989, ed. T. S. Ke (Int. Acad. Publ., Pergamon Press, 1989) p. 337.

114. R. De Batist, in M^3D, *Mechanics and Mechanisms of Material Damping*, ed. Kinra and Wolfenden (ASTM STP 1169, 1992) p. 45.

115. L. M. Clarebrough, *Acta Metall.*, **5** (1957) 413.

116. A. Chilarducci and M. Ahlers, *Scr. Metall.*, **14**, (1980) 1341.

117. A. Ghilarducci and M. Ahlers, *J. Phys. F. Met. Phys.*, **13** (1983) 1757.

118. J. Van Humbeeck and L. Delaey, *J. de Phys.*, **C9**, Suppl. n. 12, 44 (1983) 217.

119. J. E. Bidaux, R. Schaller and W. Benoit, *Acta Metall.*, **37** (1989) 803.

120. J. Stoiber, Ph.D. Thesis, Lausanne-EPFL, thèse n. 1115 (1993).

121. M. Morin, G. Guenin and P. F. Gobin, *J. de Phys.*, **43** (1982) 685.

122. W. Dejonghe, R. De Batist and L. Delaey, *Scr. Metall.*, **10** (1976).

123. R. Gotthardt and O. Mercier, *J. de Phys.*, **42** (1981) C5, 995.

124. AMT-Belgium, Commercial brochure.

125. M. Sade, R. Rapacioli and M. Ahlers, *Acta Metall.*, **33** (1985) 487.

126. M. Thumann and E. Hornbogen, *Z. Metallkd.*, **79** (1988) 119.

127. M. Sade, J. Kumpfert and E. Hornbogen, *Z. Metallkd.*, **79** (1988) 678.

128. M. Sade, E. Hornbogen, *Z. Metallkd.*, **79** (1988) 782.

129. K. N. Melton and O. Mercier, *Mater. Sci. Eng.*, **40** (1979) 81.

130. L. Delaey, J. Janssen, d. Van de Mosselaer, G. Dullenkopf and A Deruyttere, *Scr. Metall.*, **12** (1978) 373.

131. H. Sakamoto, *Trans. JIM*, **24** (1983) 665.

132. H. Sakamoto, Y. Kijima and K. Shimizu, *Trans. JIM*, **23** (1982) 585.

133. L. C. Brown, *Metall. Trans. A*, **13A** (1982) 26.

134. H. Sakamoto, K. Shimizu and K. Otsuka, *Trans. JIM*, **22** (1981) 579.

135. L. C. Brown, *Metall. Trans. A*, **10A** (1979) 217.

136. N. Y. C. Yang, C. Laird and D. P. Pope, *Metall. Trans. A*, **8** (1977) 955.

137. C. Lopez de Castillo, S. Hernaez and B. G. Mellor, *J. Mater. Sci.*, **21** (1986) 4043.

138. C. Picornell, E. Cesari and M. Sade, *Mater. Sci. Forum*, **56–58** (1990) 741.

139. S. Miyazaki and K. Otsuka, *ISIJ Int.*, **29** (1989) 353.
140. J. Janssen, F. Willems, B. Verelst, J. Maertens and L. Delaey, *Proc. ICOMAT-82* (Leuven, 1982) p. C4-809.
141. R. H. Dauskardt, T. W. Duerig and R. U. Ritchie, *MRS Proc.* (Tokyo, 1989), **9**, p. 243.
142. S. Miyazaki, Y. Sugaya and K. Otsuka, *MRS Proc.* (Tokyo, 1989) **9**, p. 251.
143. T. Tadaki, Y. Nakata and K. Shimizu, *Trans. JIM*, **28** (1987) 883.
144. J. Perkins and P. Bobowiec, *Metall. Trans. A*, **17A** (1986) 195.
145. K. Tsuji, Y. Takegawa and K. Kojima, *Mater. Sci. Eng.*, **A136** (1991) L1–L4.
146. J. L. Proft, K. N. Melton and T. W. Duerig, *MRS Proc.* (Tokyo, 1988) **9**, p. 159.
147. T. Tadaki and K. Shimizu, *MRS Proc.* (Tokyo, 1988) **9**, p. 291.
148. S. Edo, *J. of Mater. Sci. Letters*, **24** (1989) 3991.
149. Y. Furuya, H. Shimada, M. Matsumoto and T. Harma, *J. Japan Inst. Met.*, **52** (1988) 139.
150. C. M. Friend and A. P. Miodownik, *Proc. of Int. Conf. on Mart. Transf.* (1986), ed. Jap. Inst. of Metals, p. 902.
151. K. Murakami, Y. Murakami, K. Mishima and Y. Ikai, *J. Japan Inst. Metals*, **48** (1984) 115.
152. J. Beyer, B. Koopman, P. A. Besselink and P. F. Willemse, *Mater. Sci. Forum*, **56–58** (1990) 773.
153. Y. Furuya, H. Shimada, M. Matsumoto, T. Hamma, *MRS Proc.* (Tokyo, 1988) **9**, p. 269.
154. Y. Suzuki, *MRS Proc.* (Tokyo, 1988) **9**, p. 557.
155. M. Uehara, and T. Suzuki, *J. Soc. Mater. Sci. Japan*, **34** (1985) 779.
156. A. Amengual, C. Picornell, R. Rapacioli, C. Ségui and V. Torra, *Thermochim. Acta*, **145** (1989) 101.
157. R. Stalmans, J. Van Humbeeck and L. Delaey, *J. de Phys.* III, **1** (1991) C4-403.
158. P. Rodriguez and G. Guénin, *Mater. Sci. Forum*, **55–58** (1990) 541.
159. M. Mantel, Ph.D. thesis, INSA-Lyon.
160. L. Contardo, Ph.D. thesis, NX88 ISAL 0048, INSA-Lyon, 1988.
161. Y. Suzuki, *MRS-Proc.* (Tokyo, 1988) **9**, ed. MRS 1989, p. 557.
162. W. J. Moberly and K. N. Melton, in *Engineering Aspects of Shape Memory Alloys*, ed. T. W. Duerig, K. N. Melton, D. Stöckel and C. M. Wayman (Butterworth–Heinemann, London, 1990) p. 46.
163. P. G. Lindquist and C. M. Wayman, in *Engineering Aspects of Shape Memory Alloys*, ed. T. W. Duerig, K. N. Melton, D. Stöckel and C. M. Wayman (Butterworth–Heinemann, London, 1990) p. 58.
164. M. H. Wu, in *Engineering Aspects of Shape Memory Alloys*, ed. T. W. Duerig, K. N. Melton, D. Stöckel and C. M. Wayman (Butterworth–Heinemann, London, 1990) p. 69.
165. W. Duerig, K. N. Melton and J. L. Proft, in *Engineering Aspects of Shape Memory Alloys*, ed. T. W. Duerig, K. N. Melton, D. Stöckel and C. M. Wayman (Butterworth–Heinemann, London, 1990) p. 130.
166. T . W. Duerig, D. Stöckel and A. Keeley, in *Engineering Aspects of Shape Memory Alloys*, ed. T. W. Duerig, K. N. Melton, D. Stöckel and C. M. Wayman (Butterworth–Heinemann, London, 1990) p. 181.
167. P. Tautzenberger, in *Engineering Aspects of Shape Memory Alloys*, ed. T. W. Duerig, K. N. Melton, D. Stöckel and C. M. Wayman (Butterworth–Heinemann, London, 1990) p. 207.
168. W. Duerig and R. Zadno, in *Engineering Aspects of Shape Memory Alloys*, ed. T. W. Duerig, K. N. Melton, D. Stöckel and C. M. Wayman (Butterworth–Heinemann, London, 1990) p. 369.
169. Aernoudt, J. Van Humbeeck, L. Delaey and W. Van Moorleghem, in *The Science and Technology of Shape Memory Alloys*, ed. V. Torra (Impresrapit, Barcelona, 1987) p. 221.
170. B. G. Mellor, in *The Science and Technology of Shape Memory Alloys*, ed. V. Torra (Impresrapit, Barcelona, 1987) p. 334.
171. J. Perkins, *Metals Forum*, **4** (1981) 153.
172. P. A. Besselink, in *The Science and Technology of Shape Memory Alloys*, ed. V. Torra (Impresrapit, Barcelona, 1987) p. 407.

173. J. Van Humbeeck and L. Delaey, in *The Martensitic Transformation in Science and Technology*, ed. E. Hornbogen and N. Jost (DGM Informationsgesellschaft, 1989) p. 15.
174. T. W. Duerig and K. N. Melton, in *The Martensitic Transformation in Science and Technology*, ed. E. Hornbogen and N. Jost (DGM Informationsgesellschaft, 1989) p. 191.
175. P. Tautzenberger, in *The Martensitic Transformation in Science and Technology*, ed. E. Hornbogen and N. Jost (DGM Informationsgesellschaft, 1989) p. 213.
176. J. Van Humbeeck, M. Chandrasekaran and R. Stalmans, in *Proc. Int. Conf. on Mart. Trans. '92*, ed. C. M. Wayman and J. Perkins (Monterey Institute of Advanced Studies, Carmel, 1993) p. 1015.
177. J. H. Mulder, J. H. Maas and J. Beyer, in *Proc. Int. Conf. on Mart. Trans. '92*, ed. C. M. Wayman and J. Perkins (Monterey Institute of Advanced Studies, Carmel, 1993) p. 869.
178. H. Horikawa, Y. Suzuki, A. Horie, S. Yamamoto and Y. Yasuda, in *Proc. Int. Conf. on Mart. Transf. '92*, ed. C. M. Wayman and J. Perkins (Monterey Institute of Advanced Studies, Carmel, 1993) p. 1271.
179. AMT nv, Commercial brochure.
180. ALKOR, Commercial brochure.
181. EUROPA METALLI-LMI spa, Commercial brochure.
182. *Control Design with Vease – via Electrically Actuated Shape-Memory Effect*, Raychem, Commercial brochure.
183. J. Van Humbeeck, M. Chandrasekaran and L. Delaey, *Endeavour*, **15** (1991) 148.
184. D. Stoeckel, *Advanced Materials & Processes*, **10** (1990) 33.
185. *Shape-Memory Metal*, Raychem, Commercial brochure.
186. E. Hornbogen, *Practical Metallography*, **26** (1989) 279.
187. Weijia Tang, *Evaluation of property data for TiNi Shape Memory Alloys*, Licentiate Thesis, Report KTH-AMT-93, ISSN 0282-9770 Stockholm.

8

Shape memory ceramics

K. UCHINO

8.1 Development trends of new principle actuators

Recent development of new principle actuators aimed at replacing conventional electromagnetic motors has been remarkable in the following three areas; precision positioning, vibration suppression and miniature motors. Particular attention has been given to piezoelectric/electrostrictive ceramic actuators, shape memory devices of alloys such as Ni–Ti and Cu–Zn–Al, and magnetostrictive actuators using Terfenol D (Tb–Dy–Fe) alloys. Rigid strains induced in a piezoelectric ceramic by an external electric field have been used as ultraprecision cutting machines, in the Hubble telescope on the space shuttle and in dot-matrix printer heads.[1-5] There has also been proposed a parabolic antenna made of shape memory alloy, which is in a compactly folded shape when first launched on an artificial satellite, and subsequently recovers its original shape in space when exposed to the heat of the sun. Smart skins on submarines or military tanks are new targets for solid state actuators.[6]

In general, thermally-driven actuators such as shape memory alloys can show very large strains, but require large drive energy and exhibit slow response. Magnetic field-driven magnetostrictive devices have serious problems with size because of the necessity for the magnetic coil and shield. The subsequent Joule heat causes thermal dilatation in the system, and leakage magnetic field sometimes interferes with the operating hybrid electronic circuitry. On the contrary, electric field-driven piezoelectric and electrostrictive actuators have been most developed because of their high efficiency, quick response, compact size, no generation of heat or magnetic field, in spite of relatively small induced strains.

This chapter reviews shape memory properties of ceramics, focusing on antiferroelectric lead zirconate titanate ($Pb(Zr,Ti)O_3$, PZT) based ceramics. The shape memory effect can be observed not only in special alloys but also in ceramics such as partially stabilized zirconia and ferroelectric lead zirconate

184

titanate. A new concept of 'shape memory' is also proposed in this article: the elastic strain change associated with the electric field-induced phase transition is utilized instead of stress-induced or thermally-induced phase transitions, which enables many more smart actuator applications than the piezoelectrics or electrostrictors.

The principle of the ceramic shape memory effect is described firstly in comparison with the case of alloys. Phase diagrams, domain reversal mechanisms and fundamental actuator characteristics are then discussed, followed by the practical distinctions between these new ceramics and shape memory alloys. Finally, possible unique applications are proposed including a latching relay and a mechanical clamp.

8.2 Shape memory ceramics

Shape memory effect is observed not only in special alloys but also in ceramics or in polymers. The shape memory effect in alloys originates from a thermally-induced or stress-induced 'martensitic' phase transition. After the alloy is deformed largely in the martensitic state, this apparently permanent strain is recovered to its original shape when heated to cause the reverse martensitic transition. Then, upon cooling, the shape returns to its original state (see Fig. 8.1(a)).

A similar effect is anticipated in ceramics with a certain phase transition, i.e. a 'ferroelastic' phase transition. Reyes-Morel *et al.* demonstrated the shape memory effect as well as superelasticity in a CeO_2-stabilized tetragonal zirconia (ZrO_2) polycrystal.[7] Figure 8.2 is cited from their data, which shows the uniaxial compressive stress versus strain curve at room temperature, together with the temperature–strain curve showing strain recovery on heating. Under uniaxial compression, the specimen deforms plastically owing to a stress-induced tetragonal to monoclinic transition in Ce-doped zirconia. Continuous deformation is interrupted by repeated load drops, providing a nearly constant upper yield stress of 0.7 GPa. Even after unloading, large residual plastic axial strain (-0.7%) is observed. Subsequent heating produces a gradual recovery of the residual strain due to the reverse phase transition starting at 60 °C and a burst of strain recovery at 186 °C. The burst is very sharp, above which approximately 95% of the prior axial strain is recovered.

Ceramics 'shape memory' has been reported also for certain ferroelectricity-related transitions, namely paraelectric–ferroelectric[8] and antiferroelectric–ferroelectric transitions.[9,10] The former thermally-induced transition revealed a shape-recovery phenomenon similar to zirconia ceramics. On the contrary, the latter is related to an electric field-induced transition, and exhibits large

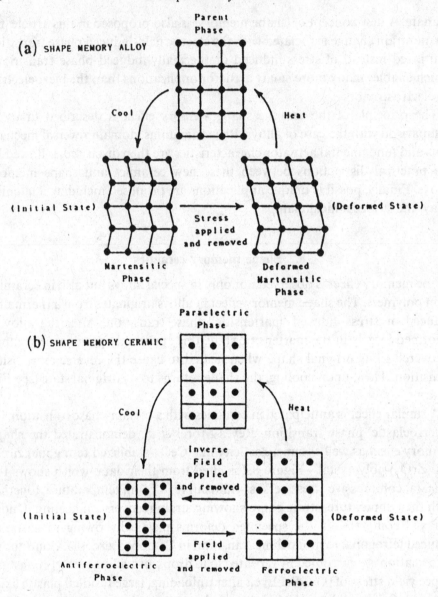

Fig. 8.1. Comparison of the mechanisms for the shape memory effect in alloys and in antiferroelectric ceramics.

displacement (0.4%) with a 'digital' characteristic or a shape memory function, which is in contrast to the essentially 'analogue' nature of conventional piezoelectric/electrostrictive strains with 0.1% in magnitude.

Let us review ferroelectricity and antiferroelectricity here for further understanding.[11] Figure 8.3 shows the crystal structure changes in a typical fer-

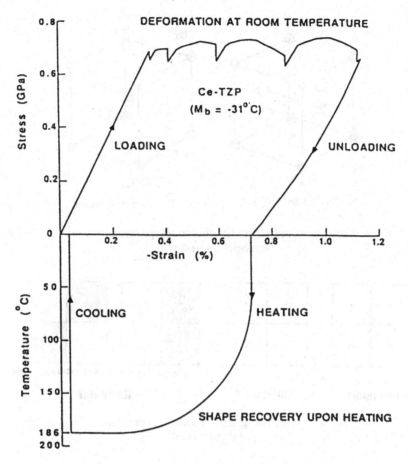

Fig. 8.2. Uniaxial compressive stress *vs* strain curve for Ce-doped tetragonal zirconia polycrystal at room temperature, together with temperature–strain curve showing strain recovery. (After Reyes-Morel *et al.*[7])

roelectric, barium titanate ($BaTiO_3$, BT). At an elevated temperature above the transition point of 130°C (Curie temperature, denoted as T_C), BT shows a cubic 'perovskite' structure (paraelectric (PE) phase) as illustrated in Fig. 8.3(a). With decreasing temperature below T_C, the cations (Ba^{2+} and Ti^{4+}) shift against the anions (O^{2-}) as illustrated in Fig. 8.3(b), exhibiting spontaneous polarization as well as spontaneous strain (ferroelectric (FE) phase). Notice that the electric dipole moments in each crystal unit cell are arranged in parallel in a ferroelectric. On the other hand, there exists an antiferroelectric where the dipoles are arranged antiparallel to each other so as not to produce net polarization. Figure 8.4 shows two antipolar dipole arrangement models in contrast to nonpolar and polar models.

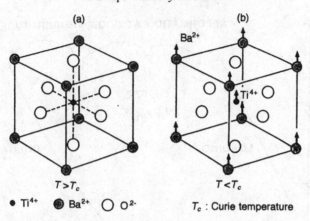

Fig. 8.3. Crystal structures of barium titanate in the paraelectric state (a), and in the ferroelectric state (b).

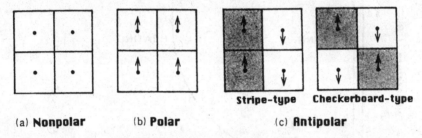

Fig. 8.4. Two antipolar dipole arrangement models (c) in contrast to nonpolar (a) and polar (b) models.

When the free energy of the antipolar state is close to the energy of the polar state, the dipole configuration is rearranged by the external electric field or stress. Figure 8.5 shows the applied electric field versus induced polarization curves in PE, FE and AFE materials. A linear relation and a hysteresis due to the spontaneous polarization reversal between positive and negative directions are observed in a PE and in a FE, respectively. On the contrary, an AFE exhibits an electric field-induced phase transition to a FE state above a critical field E_t, accompanied by a hysteresis above E_t. Reducing the field down to zero, the remanent polarization is not observed, providing a so-called 'double hysteresis' curve in total. A large strain jump is theoretically associated with this phase transition, which also appears as a double hysteresis. In a certain case, once the FE state is induced, this FE state is sustained even if the electric field is decreased to zero; this corresponds to the 'shape memory' phenomenon. The mechanism for the shape memory effect in the AFE ceramics is schematically illustrated in Fig. 8.1(b).

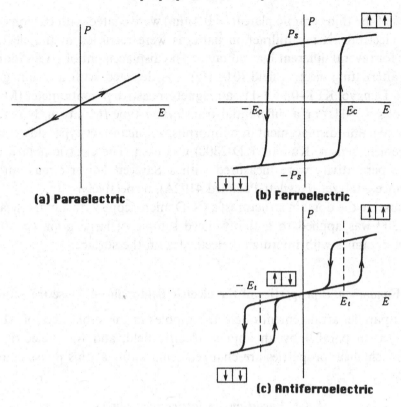

Fig. 8.5. Electric field *vs* induced polarization curves in paraelectric (a), ferroelectric (b) and antiferroelectric (c) materials.

8.3 Sample preparation and experiments

This section introduces sample preparation and experimental procedures of shape memory ceramics.

Antiferroelectric perovskite ceramics from the PZT system have been investigated in which successive phase transitions from a PE, through an AFE, to a FE state appear with decreasing temperature.[12] PZT ceramics $Pb_{0.99}Nb_{0.02}$ $[(Zr_{0.6}Sn_{0.4})_{1-y}Ti_y)]_{0.98}O_3$ ($0.05 < y < 0.09$) (abbreviated hereafter as PNZST) were prepared from reagent grade oxide raw materials, PbO, Nb_2O_5, ZrO_2, SnO_2 and TiO_2. Bulk samples were prepared by hot-press sintering at 1200°C. Unimorphs were fabricated with two thin rectangular plates (22 mm × 7 mm × 0.2 mmt) bonded together. Multilayer samples (12 mm × 4.3 mm × 4.3 mmt) with each layer 140 μm in thickness were also fabricated using the tape casting technique: those with platinum electrodes were sintered at a temperature of about 1250°C.

The field-induced lattice change was determined by x-ray diffraction. The surfaces of the thin ceramic plate ($t = 0.2\,\text{mm}$) were coated with carbon evap-orated electrode. X-ray diffraction patterns were recorded at the electrode surface for several different bias voltages. The displacement or strain induced by an alternating electric field ($0.05\,\text{Hz}$) was detected with a strain gauge (Kyowa Dengyo, KFR-02-C1-11), a magneto-resistive potentiometer (Midori Precisions, LP-1U) or a differential transformer-type (Millitron, No. 1202). For the dynamic displacement in unimorphs, a noncontact-type eddy current displacement sensor (Kaman, KD-2300) was used. The electric polarization and the permittivity were measured with a Sawyer–Tower circuit and an impedance analyzer (Hewlett-Packard 4192A), respectively.

To observe the domain structures, a CCD microscope with a magnification of $\times 1300$ was applied to a thinly-sliced sample of large grain ($> 50\,\mu\text{m}$) PNZST ceramics with interdigital electrodes on the surface.

8.4 Fundamental properties of the electric field-induced phase transition

The antiparallel arrangement of electric dipoles in the sublattices of AFE is rearranged in parallel by an applied electric field, and the dielectric and electromechanical properties are changed remarkably at this phase change.

8.4.1 Variations in lattice parameters

The field-induced change in lattice parameters for a sample $Pb_{0.99}Nb_{0.02}$ $[(Zr_{0.6}Sn_{0.4})_{1-y}\,Ti_y]_{0.98}O_3$ with $y = 0.06$ is plotted in Fig. 8.6(a).[13] The forced transition from the AFE to the FE phase gives rise to the simultaneous increase of a and c in the perovskite unit cell, thereby keeping the tetragonality, c/a, nearly constant. Since y makes only a negligible contribution to the volume change, the strain change at the phase transition is nearly isotropic with a magnitude of $\Delta L/L = 8.5 \times 10^{-4}$.

The intensity change of the x-ray reflections with the application of an electric field suggests that the spontaneous polarization in the FE state lies in the c-plane, parallel with the perovskite [110] axis, and that the sublattice polarization configuration in the AFE state is very similar to that of $PbZrO_3$.[14] Figure 8.6(b) illustrates the simplest two-sublattice model.

8.4.2 Temperature dependence of the induced strain

The temperature dependence of the field-induced strain is described for the sample with $y = 0.06$ in conjunction with the dielectric measurements.[13]

Figure 8.7 shows the relation between the electric field and polarization. The

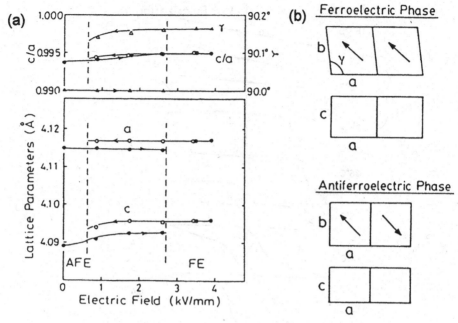

Fig. 8.6. (a) Variation of lattice parameters with bias electric field at room temperature ($y = 0.06$). (b) Two-sublattice model of the polarization configuration for the AFE and FE states. (After Uchino and Nomura[13])

typical double and ferroelectric hysteresis loops are observed at room temperature and $-76°C$, respectively, while a transitive shape with humps is observed at intermediate temperatures.

The transitive process can be observed more clearly in the strain curve. Figure 8.8 shows the transversely induced strains. The forced transition from AFE to FE at room temperature is characterized by a huge strain discontinuity. On the other hand, a typical ferroelectric butterfly-type hysteresis is observed at $-76°C$, corresponding to polarization reversal. It is important to note that the strain discontinuities associated with the phase transition have the same positive expansion in both longitudinal and transverse directions with respect to the electric field (i.e. the apparent Poisson's ratio is negative!), while the piezostriction is negative and positive in the transverse and longitudinal directions, respectively.

The shape memory effect is observed on this loop at $-4°C$. When a large electric field is applied to the annealed AFE sample, a massive strain $\Delta L/L$ of about 7×10^{-4} is produced and maintained metastably even after the field is removed. After applying a small reverse field or thermal annealing, the original AFE shape is observed.

The reverse critical field related to the FE–AFE transition is plotted with

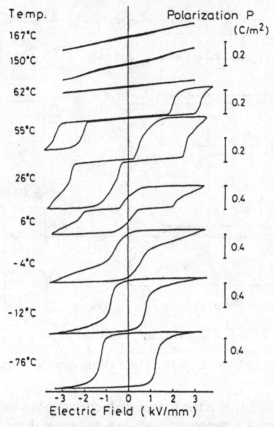

Fig. 8.7. Polarization plotted as a function of electric field for several temperatures
($y = 0.06$). (After Uchino and Nomura[13])

solid linès in the phase diagram for the sample with $y = 0.06$ (see Fig. 8.9). In
the temperature range from $-30°C$ to $10°C$, a hump-type hysteresis in the
field versus polarization curve and an inverse hysteresis in the field-induced
strain are observed: this has previously often been misinterpreted as another
AFE phase different from the phase above $10°C$. The annealed state below
$-30°C$ down to $-200°C$ is AFE. However, once the FE state is induced, the
AFE phase is never observed during a cycle of rising and falling electric field.
The critical field line for the FE to AFE transition (the solid line) in the
temperature range $-30°C$ to $10°C$ intersects the coercive field line for the
$+$ FE to $-$ FE reversal (the dashed line) below $-30°C$.

8.4.3 Composition dependence of the induced strain

Figure 8.10 shows the strain curves induced transversely by the external field at
room temperature for samples of several different compositions.[15] The molar

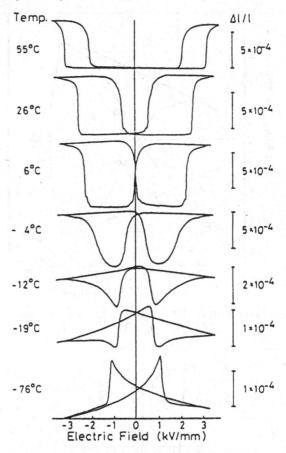

Fig. 8.8. Transverse elastic strain induced by the electric field at several temperatures
($y = 0.06$). (After Uchino and Nomura[13])

fraction of Ti, y, is increased from 0.06 (a) to 0.065 (c). The initial state was
obtained by annealing at 150°C, which is above the Curie (or Néel) temperature for all the samples. A typical double hysteresis curve (Type I) is observed
in the sample containing $y = 0.06$. Large jumps in the strain are observed at
the forced phase transitions from the AFE to the FE phase ($\Delta L/L = 8 \times 10^{-4}$).
In comparison, the strain change with electric field in either the AFE or FE
state is rather small: this suggests a possible application for the material as a
'digital' displacement transducer, having OFF/ON displacive states. The difference in the strain between that occurring in the initial state and that
appearing in a cyclic process at $E = 0\,kV/cm$ is also noteworthy and will be
explained in the following section.

In the sample with $y = 0.063$, a Ti concentration slightly higher than that
just described, the field-induced FE phase will not return to the AFE state even
after decreasing the field to zero (Type II, Fig. 8.10(b)): this is called 'memor-

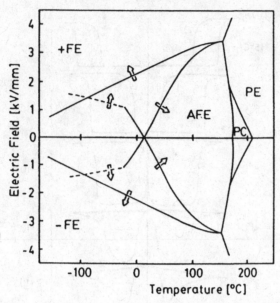

Fig. 8.9. Phase diagram on the temperature–electric field plane for $Pb_{0.99}Nb_{0.02}$ $[(Zr_{0.6}Sn_{0.4})_{0.94}Ti_{0.06}]_{0.98}O_3$.

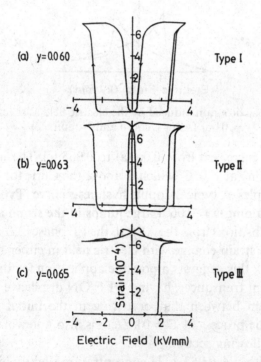

Fig. 8.10. Transverse induced strains of $Pb_{0.99}Nb_{0.02}$ $[(Zr_{0.6}Sn_{0.4})_{1-y}Ti_y]_{0.98}O_3$ at room temperature: (a) for $y = 0.06$; (b) for $y = 0.063$; and (c) for $y = 0.065$. (After Uchino[15])

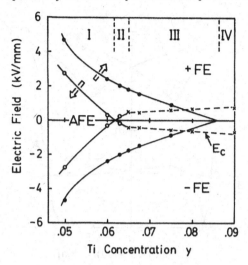

Fig. 8.11. Phase diagram of $Pb_{0.99}Nb_{0.02}[(Zr_{0.6}Sn_{0.4})_{1-y}Ti_y]_{0.98}O_3$ at room temperature with respect to the composition y and the applied electric field E.

izing' the FE strain state. In order to obtain the initial AFE state, a small reverse bias field is required. Figure 8.10(c) shows the strain curve for the sample with $y = 0.065$, which exhibits irreversible characteristics during an electric field cycle (Type III). The initial strain state can only be recovered by thermal annealing up to 50°C.

Data derived from these strain curves may be utilized to construct a phase diagram of the system $Pb_{0.99}Nb_{0.02}[(Zr_{0.6}Sn_{0.4})_{1-y}Ti_y]_{0.98}O_3$ at room temperature with respect to the composition y and the applied electric field E (Fig. 8.11). If the Ti concentration of the horizontal axis is redefined in terms of temperature and evaluated in the opposite direction, this phase diagram is topologically the same as the phase diagram of Fig. 8.9. The key feature of this phase diagram is the existence of the three phases, namely the AFE, the positively poled FE ($+$ FE), and the negatively poled FE ($-$ FE) phases, the boundaries of which are characterized by the two transition lines corresponding to rising and falling electric fields.

The composition regions I and IV exhibit the typical double hysteresis and ferroelectric domain reversal, respectively. The shape memory effect is observed in regions II and III. It is important to consider the magnitude of the electric field associated with the $+$ FE → AFE transition (notice the direction of the arrow!). Let us consider the transition process under an inverse bias field after the $+$ FE is induced by the positive electric field. If the magnitude of the field for the $+$ FE → AFE transition is smaller than the coercive field for $+$ FE → $-$ FE (region II, $0.0625 < y < 0.065$), the AFE phase appears once

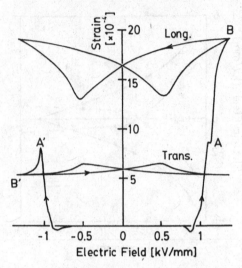

Fig. 8.12. Longitudinal and transverse induced strains of the sample $Pb_{0.99}Nb_{0.02}$ $[(Zr_{0.6}Sn_{0.4})_{0.925}Ti_{0.075}]_{0.98}O_3$. (After Uchino[15])

under a small inverse field, then the $-$ FE phase is induced at the AFE $\rightarrow$ $-$ FE transition field. In this case, the shape memory is reversible to the initial state only with the application of a reverse electric field (Type II): this is very useful! On the other hand, if the $+$ FE $\rightarrow$ $-$ FE coercive field is smaller than the $+$ FE $\rightarrow$ AFE field (region III, $0.0625 < y < 0.085$), the domain reversal to $-$ FE appears without passing through the AFE phase. The initial state can be obtained by thermally annealing up to 50–70°C (Type III).

8.4.4 Domain reorientation mechanism in antiferroelectrics

Antiferroelectrics cannot be poled macroscopically. However, since they have sublattice polarizations closely coupled with the lattice distortion, it is possible to consider ferroelastic domain orientations in antiferroelectrics. This is a possible approach to understanding the difference in the strain between that occurring in the initial state and that arising in a cyclic process, as shown in Fig. 8.10(a).

Figure 8.12 shows the longitudinal and transverse strains induced in the sample $y = 0.075$ in region III.[15] The strain induction process can be considered to consist of two stages: first, there is an isotropic volume expansion (O $\rightarrow$ A, A': $\Delta L/L = 8 \times 10^{-4}$) due to the AFE to FE phase transition (remember Fig. 8.6(a), where the perovskite cell expands isotropically by $\Delta L/L = 8.5 \times 10^{-4}$), and second, there is an anisotropic strain associated with the FE domain rotation (A $\rightarrow$ B, A' $\rightarrow$ B': $x_3 = 9 \times 10^{-4}$, $x_1 = -3 \times 10^{-4}$). This process is

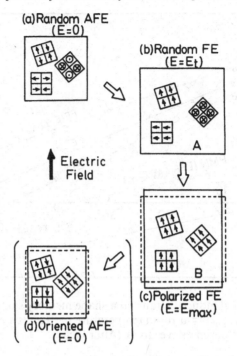

Fig. 8.13. Schematic illustration of the antiferroelectric domain reorientation associated with the forced phase transition.

shown schematically in Fig. 8.13, where a probable model for the double-
hysteresis sample (region I) is also illustrated. As previously pointed out, even
for AFEs, domain reorientation is possible through the forced phase transition
to FEs.

Domain configuration in PNZST $y = 0.063$ (Type II) was observed as a
function of electric field at room temperature.[16] No clear domains were
observed in the initial state obtained by annealing the sample at 70°C. As the
electric field increased, clear domain walls appeared above 20 kV/cm, arranged
almost perpendicularly to the electric field direction. This value of electric field
is coincident with the critical field which can cause the transition from AFE to
FE. Therefore, these domain walls were caused by the induced ferroelectricity.
The domain walls did not diminish whilst removing the electric field, because
the sample has the shape memory effect. The walls disappeared when slightly
negative bias was applied, as expected in the Type II specimen.

8.4.5 Pressure dependence of the field-induced strain

One of the most important criteria for an actuator is reliable and stable driving
under a large applied stress, as required for its application as a positioner in

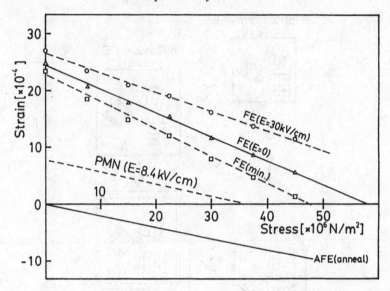

Fig. 8.14. Longitudinal induced strains in a shape memory sample $Pb_{0.99}Nb_{0.02}$ $[(Zr_{0.6}Sn_{0.4})_{0.93}Ti_{0.07}]_{0.98}O_3$ at room temperature plotted as a function of uniaxial compressive stress. (After Uchino[10])

precision cutting machinery. Figure 8.14 shows the longitudinally induced strains in a shape memory sample of PNZST $y = 0.0707$ in both the AFE and FE states at room temperature plotted as a function of uniaxial compressive stress for several electric fields.[10] For comparison, similar plots for a lead magnesium niobate (PMN) based ceramic $(Pb(Mg_{1/3}Nb_{2/3})_{0.65}Ti_{0.35}O_3$, a well-known electrostrictive material) are also shown.

Roughly speaking, the strain versus stress curve for the AFE PNZST ceramic is shifted along the strain axis with respect to that for PMN due to the difference between the spontaneous strains in the AFE and the FE states. Consequently, the maximum generative force obtained when the ceramic is mechanically clamped so as not to generate a displacement is raised to 80 MPa in comparison with the normal value for the ferroelectric, 35 MPa.

8.5 Comparison with shape memory alloys

The phenomenon associated with shape memory alloys is attributed to the stress-induced (as well as thermally-induced) phase transition referred to as 'martensitic'. The new strain phenomena in AFE ceramics described here are very easily understandable, if we use the terminology conventionally used for these alloys, replacing electric field E for stress X. The 'digital displacement'

Table 8.1. *Comparison of the shape memory characteristics between alloys and antiferroelectric ceramics*

Properties	Antiferroelectric	Shape memory alloy
Driving power	Voltage (mW $\sim$ W)	Heat (W $\sim$ kW)
Strain ($\Delta L/L$)	$10^{-3} \sim 10^{-2}$	$10^{-2} \sim 10^{-1}$
Generative force	100 MPa	1000 MPa
Response speed	msec	sec $\sim$ min
Durability	$> 10^6$ cycles	10^4 cycles

and the ferroelectric-state memorization discussed here correspond to the 'superelasticity' and the shape memory effect in the alloys, respectively.

Outstanding merits of the ceramics over the alloys are:

(1) quick response in msec,
(2) good controllability by electric field to memorize and recover the shape without generating heat,
(3) low energy consumption as low as 1/100 of the alloy, and
(4) wide space is not required to obtain the initial shape deformation.

Numerical comparison is shown in Table 8.1.

8.6 Applications of shape memory ceramics

The conventional piezoelectric/electrostrictive actuators have been developed with the aim of realizing 'analogue displacement transducers', in which a certain magnitude of electric field corresponds to only one strain state without any hysteresis during rising and falling electric field. This is exemplified by the electrostrictive PMN based ceramics. On the contrary, the antiferroelectrics introduced in this article may be utilized in a device based on a new concept, 'a digital displacement transducer', in which bistable ON/OFF strain states exist for a certain electric field. This idea may be interpreted as a stepping motor in the conventional terminology of electromagnetic motors. The discrete movement through a constant distance achieved by the new actuator is well suited for applications such as an optical-grid manufacturing apparatus or a swing-type charge coupled device.

The shape memory material can be applied to devices such as latching relays and mechanical clampers, where the ceramic is capable of maintaining the excited ON state even when electricity is not applied to it.

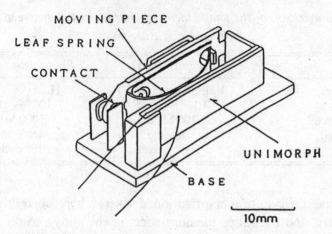

Fig. 8.15. Structure of the latching relay using shape memory ceramic.

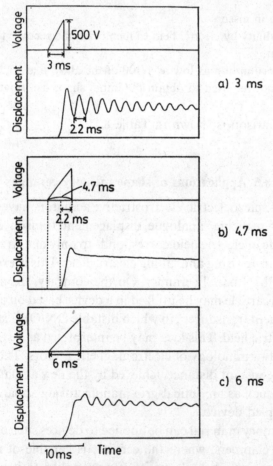

Fig. 8.16. Dynamical response of the tip displacement of the shape memory unimorph
with $y = 0.063$ under various drive pulse conditions. (After Furata et al.[17])

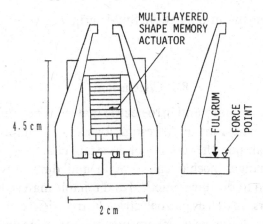

Fig. 8.17. Construction of the mechanical clamper using a shape memory multilayer device.

8.6.1 Latching relay

Figure 8.15 shows the structure of a newly fabricated latching relay, which is composed essentially of a mechanical snap-action switch and a shape memory unimorph driving part.[17] The snap-action switch is easily driven by a $50 \mu m$ displacement, having mechanically bistable states. The unimorph is fabricated with two $y = 0.063$ ceramic plates of $22 mm \times 7 mm$ area and $0.2 mm$ thickness, bonded together with adhesive.

Figure 8.16 shows the dynamical response of the unimorph. It is important to note that the phase transition arises quickly enough to generate the following mechanical resonant vibration (Fig. 8.16(a)). When the rise time of the electric field is adjusted to $4.7 ms$ (Fig. 8.16(b)), which is the sum of the mechanical resonant period ($2.2 ms$) and the lag time to cause the phase transition ($2.5 ms$), the ringing can be suppressed completely.

The new relay is very compact in size, 1/10 of a conventional electromagnetic type, and is operated by a pulse voltage, which provides a significant energy saving. The relay is turned ON at $350 V$ with $4 ms$ rise time, and turned OFF at $-50 V$.

8.6.2 Mechanical clamper

A mechanical clamper suitable for microscope sample holders has been constructed by combining a 20-layer shape memory stacked device ($y = 0.0635$) and a hinge-lever mechanism as shown in Fig. 8.17.[18] Application of a $1 ms$ pulse voltage of $200 V$ can generate the longitudinal displacement of $4 \mu m$ in the $4 mm$-thick multilayer device, leading to $30 \mu m$ tip movement of the hinge lever after displacement amplification. Stable grip was verified for more than several hours.

8.7 Conclusions

The study of shape memory antiferroelectric materials has only just begun in the past several years. Further investigations on the improvement of the induced strain magnitude, the stability of the strain characteristics with respect to temperature change, mechanical strength and durability after repeated driving are required to produce practical and reliable materials. This category of ceramic actuators, as well as piezoelectric/electrostrictive materials, will be a vital new element in the next generation of 'micro-mechatronic' or electromechanical actuator devices.

References

1. H. Yamada, ed., *Essentials to Developments and Applications of New Method/Principle Motors*, (Jpn. Industrial Tech. Center, Tokyo, 1984).
2. K. Uchino, *Piezoelectric/Electrostrictive Actuators*, (Morikita Pub. Co., Tokyo, 1986).
3. K. Uchino, FC Annual Report for Overseas Readers, *Fine Ceram. Soc. Jpn.*, **23** (1988).
4. K. Uchino, *Mater. Res. Soc. Bull.*, **18** (4) (1993) 42.
5. K. Uchino, *Piezoelectric Actuators and Ultrasonic Motors*, (Kluwer Academic Publishers, Massachusetts, 1997).
6. K. Uchino, Chapter 1, Introduction in *Handbook on New Actuators for Precision Control*, (Fuji Technosystem, Tokyo, 1994).
7. P. E. Reyes-Morel, J. S. Cherng and I. W. Chen, *J. Amer. Ceram. Soc.*, **71** (8) (1988) 648.
8. T. Kimura, R. E. Newnham and L. E. Cross, *Phase Transitions*, **2** (1981) 113.
9. K. Uchino, *Oyo Butsuri*, **54** (6) (1985) 591.
10. K. Uchino, *Proc. MRS Int'l Mtg. on Adv. Mater.*, **9** (1989) 489.
11. K. Uchino, Chapter 12, Ferroelectric Ceramics, in *Structure and Properties of Ceramics*, ed. M. Swain, (VCH, New York, 1994).
12. D. Berlincourt, H. H. A. Krueger and B. Jaffe, *J. Phys. & Chem. Solids*, **25** (1964) 659.
13. K. Uchino and S. Nomura, *Ferroelectrics*, **50** (1983) 191.
14. H. Fujishita and S. Hoshino, *J. Phys. Soc. Jpn.*, **53** (1984) 226.
15. K. Uchino, *Jpn. J. Appl. Phys.*, **24** (24-2) (1985) 460.
16. K. Y. Oh, L. E. Cross and K. Uchino, *J. Adv. Performance Mater.* (1996) [in press].
17. A. Furata, K. Y. Oh and K. Uchino, *Sensors and Mater.*, **3** (4) (1992) 205
18. A. Furata, K. Y. Oh and K. Uchino, *Proc. Int'l Symp. Appl. Ferroelectrics '90*, (1991) 528.

9

Shape memory polymers

M. IRIE

9.1 Shape memory effect of polymer materials

The term 'polymer' refers to a large molecule made by the addition of many
molecular units, monomers. The monomers chemically bond together to make
a large molecular weight polymer.

$$nA \longrightarrow +(A)_n$$
$$\text{monomers} \qquad \text{polymer}$$

In some polymers, the fundamental monomer units are not all the same but are
two or more similar molecules. The polymer is named a copolymer.

$$nA + mC \longrightarrow +(A-C-A-C-C)_{n+m}$$
$$\text{monomers} \qquad \text{copolymer}$$

Certain polymers show shape memory effects. The mechanism is, however,
completely different from that of metal alloys. A typical shape memory polymer is represented by rubber. Rubber expands several times under stress, while
it returns immediately to the original length when the stress is removed.
Rubber shows an elastic property at room temperature. However, at low
temperature, for example at $-196\,°C$, the elasticity is lost. When the expanded
rubber is cooled at $-196\,°C$, the elongated shape is fixed and never returns to
the original shape so long as the temperature is kept below the glass transition
temperature, T_g. T_g is defined as a temperature above which the elasticity of a
polymer dramatically decreases and a stiff polymer shows rubberlike behavior
as shown in Fig. 9.1. At T_g the frozen polymer chain segments become mobile
and the slope of the volume vs temperature curve changes. When the temperature is increased above T_g, the strain is released and the deformed shape reverts
again to the original one. In rubber, a three dimensional polymer network

203

Table 9.1. *Physical properties of shape memory polymers and metal alloys*

Physical property	Polymers	Metal alloys
Density (g/cm³)	0.9–1.1	6–8
Deformation (%)	250–800	6–7
Recovery temperature (°C)	25–90	− 10–100
Force required for deformation (kgf/cm²)	10–30	500–2000
Recovery stress (kgf/cm²)	10–30	1500–3000

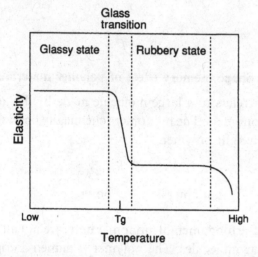

Fig. 9.1. Temperature dependence of elasticity of organic polymers.

memorizes the original shape and glassy state interchain interactions fix the shape in the transient form.

Organic shape memory polymers have the advantages of light weight, low cost, easy control of recovery temperature, and color variation. The shape memory effect can be controlled not only by heating but also by light or chemicals. Although the polymers have such advantages, their practical use is still limited. The main reason is that the polymers lack recovery stress. The stress is much less ($\sim 1/100$) than that of metal alloys. Table 9.1 compares various physical properties of shape memory polymers and metal alloys.

The mechanisms of the shape memory effects of polymers are classified as follows.

Case 1

The original shape is formed from the polymer powder or pellets by melt molding. If necessary, crosslinking is performed by the addition of crosslinking agents or by radiation. Then the shape is deformed under stress at a temperature near or above T_g or the melting temperature, T_m. The deformed shape is fixed by cooling below T_g or T_m. The deformed form reverts to the original memorized shape by heating the sample above T_g or T_m.

polymer powder, pellets $\longrightarrow$ L_0 $\longrightarrow$ $L_0 + \Delta L$
crosslinking $\quad\quad T > T_g, T_m$
melt molding

$\longrightarrow$ $L_0 + \Delta L$ $\longrightarrow$ L_0
$T < T_g, T_m$ $\quad\quad\quad T > T_g, T_m$

Case 2

Polymer shape is reversibly controlled by photo or electrochemical reactions of the polymer. The original shape is formed by crosslinking the polymer chains as in case 1.

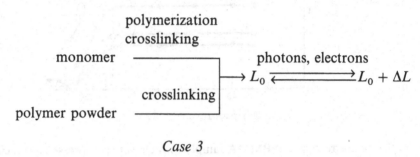

Case 3

Polymer shape is reversibly controlled by chemical reactions of the polymer.

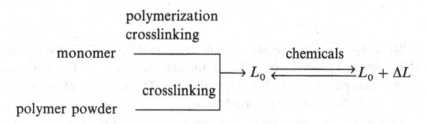

Before giving the details of the above three cases, it is worthwhile to note that widely used polymers, such as poly(methyl methacrylate), PMMA, or poly(vinyl chloride), also show shape memory effects under certain conditions. PMMA, for example, flows in the absence of crosslinking agents and cannot maintain the shape when the temperature is increased far above T_g. The shape of the polymer, however, can be changed at will under stress near T_g (below $T_g + 10°C$) and returns to the initial shape above T_g.

Figure 9.2 shows the shape recovery of two PMMA samples.[1] One sample was compressed as much as 50% at 80°C, which is 25°C below T_g, and the

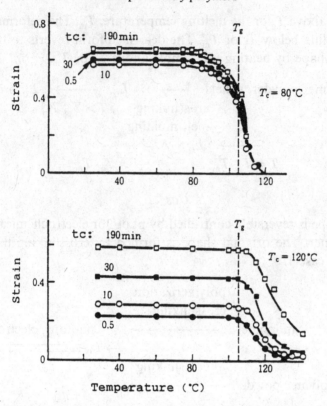

Fig. 9.2. Shape recovery of PMMA samples, which were compressed as much as 50% at 80°C and 120°C. T_c and t_c are the temperature at which the compression was carried out, and the treating time, respectively.[1]

other at 120°C. The PMMA sample treated at 80°C returned to the original shape above T_g. On the other hand, the sample treated at 120°C for more than 30 min cannot recover to the original shape. The memory of the original shape is lost when the sample is treated above $T_g + 15$°C, while the memory remains so long as the sample is treated below T_g. This kind of behavior is common in organic polymers.

9.2 Thermal-responsive shape memory effect

Table 9.2 shows various types of shape memory polymers along with physical interactions used for memorizing the original and transient shapes of the polymers. In general, interpolymer chain interactions are so weak that one dimensional polymer chains cannot keep a certain shape above T_g. To main-

Table 9.2. *Shape memory polymers and mechanisms*

Interchain interaction	Poly-norbornene	*trans-*Polyisoprene	Styrene–butadiene copolymer	Poly-urethane	Poly-ethylene
Entanglement	O				
Crosslinking		O			O
Micro-crystals		T	O, T	O	T
Glassy state	T			T	

O: used for memorizing the original shape
T: used for maintaining the transient shape

tain a stable shape, polymer chains should have a three dimensional network. Interpolymer chain interactions useful for constructing the polymer network are crystal, aggregate or glassy state formation, chemical crosslinking, and chain entanglement. The latter two interactions are permanent and used for constructing the original shape. The other interactions are thermally reversible and used for maintaining the transient shapes. In the following, five examples are described in detail.

Polynorbornene[2,3]

The chemical structure of the polymer is as follows. The molecular weight is as high as 3 million.

The long polymer chains entangle each other and a three dimensional network is formed. The polymer network keeps the original shape even above T_g in the absence of stress. Under stress the shape is deformed and the deformed shape is fixed when cooled below T_g. As already mentioned, organic polymers have glass transition temperatures, above which the polymers show rubberlike behavior. The shape can be changed at will. In the glassy state the strain is frozen and the deformed shape is fixed. The decrease in the mobility of polymer chains in the glassy state maintains the transient shape in polynorbornene. Table 9.3 shows the physical properties of polynorbornene. The disadvantage of this polymer is the difficulty of processing because of its high molecular weight.

Table 9.3. *Physical properties of shape memory polymers*

Property	Poly-norbornene	*trans-*Polyisoprene	Styrene–butadiene copolymer
Deformation (%)	~ 200	~ 400	~ 400
Recovery temperature (°C)	38	60–90	60–90
Recovery stress (kgf/cm^2)	—	10–30	5–15
Tensile strength (kgf/cm^2)	350	250	100

Polyisoprene[4,5]

Polyisoprene has the following four micro-structures.

Cis - IP

Trans - IP

Among them, *trans*-form polyisoprene (TIP) has a T_m of 67 °C and the degree of crystallinity is around 40%. The polymer can be chemically crosslinked by peroxides. Below T_m the TIP polymer has a chain network, which is connected by both chemical bonds and micro-crystal parts of the polymer. Above T_m the micro-crystal phase disappears and only the chemical bonds remain as the crosslinking points. The polymer has elasticity similar to rubber at temperatures above T_m.

Shape memory behavior of the polymer is illustrated in Fig. 9.3. The original shape is formed by heating the polymer powder or pellets with crosslinking agents at around 145 °C for 30 min and then by cooling it to room temperature. The polymer network made of the polymer chains and crosslinking agents constructs the original shape. Subsequent deformation can be done by heating the polymer at around 80 °C. The transient shape is fixed by micro-crystal parts, which are formed in the polymer during the cooling process. The deformed shape again returns to the original one upon heating above 80 °C.

The TIP polymer has relatively large recovery or shrinking stress (10–30 kgf/

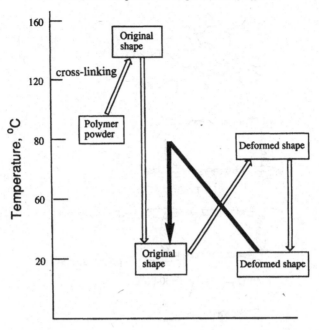

Fig. 9.3. A schematic illustration of the shape memory effect of *trans*-polyisoprene.

cm^2). The stress increases in proportion to the stretching ratio. It is possible to control the stress by changing the stretching ratio (200–400%). The disadvantage of the polymer is lack of durability because of the presence of reactive diene groups in the polymer main chain.

Styrene–butadiene copolymer[6,7]

This polymer is a copolymer. It is made of polystyrene and polybutadiene parts and these two parts are connected to one chain.

$$\left(CH_2-CH\right)_n\left(CH=CH-CH=CH\right)_m$$

The shape memory behavior of this copolymer is illustrated in Fig. 9.4. Above 120°C the copolymer melts and flows. Below 120°C polystyrene parts aggregate with each other to form a certain initial shape. Aggregate or glassy state formation in polystyrene parts is used to memorize the original shape. The deformation is carried out by heating the sample at around 80°C, at which temperature polystyrene parts are rigid but polybutadiene parts are flexible. Below 40°C the polybutadiene parts become crystallized and the deformed

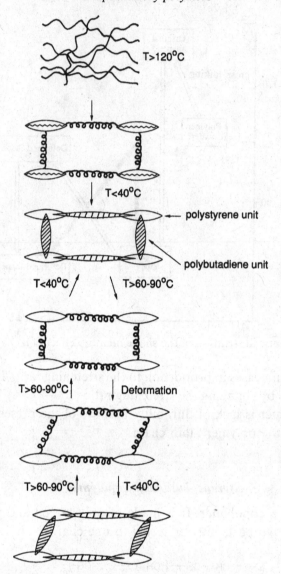

Fig. 9.4. A schematic illustration of the shape memory effect of styrene–butadiene copolymer.

shape is fixed. The shape again returns to the original one upon heating at around 80°C, at which the micro-crystals in the polybutadiene parts melt. Micro-crystal formation in the polybutadiene parts is used to fix the transient deformed shape. The crystals act as the crosslinking points of the polymer network. The physical properties of the copolymer are also summarized in Table 9.3.

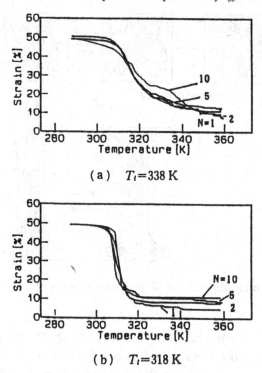

(a) $T_i = 338$ K

(b) $T_i = 318$ K

Fig. 9.5. Shape recovery of polyurethane samples, which were stretched as much as 50% at 338 and 318 K.[8]

Polyurethane[8]

Figure 9.5 shows the cycle number (N) dependence of two polyurethane samples. One sample is stretched as much as 50% at 338 K, which is 20°C higher than T_g, and the other at T_g of 318 K. The sample treated at T_g has almost returned to the original shape even after 10 cycles when heated above T_g. On the other hand, the sample treated at 338 K slowly and uncompletely returned to the original shape. The memory of the shape is lost when stretched at $T_g + 20$°C. This behavior is very similar to that of PMMA shown in Fig. 9.2.

Polyethylene[9]

Shape memory polymers are widely used for heat-shrinkable tubes. Most of them are made of crystalline polymers, such as polyethylene. Figure 9.6 illustrates the shape memory behavior. The initial shape is given by radiation-induced or chemical crosslinking. At room temperature both permanent chemical crosslinking points and micro-crystalline parts coexist. Upon heating above T_m of the polymer (around 80°C), the micro-crystalline parts melt and

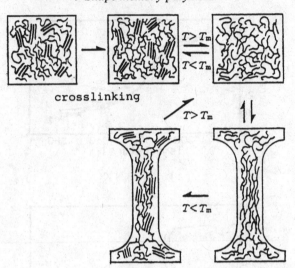

crosslinking

Fig. 9.6. A schematic illustration of the shape memory effect of polyethylene.

the polymer can be stretched under stress. At that temperature chemical crosslinking points still remain and prevent the flow of the polymer. The stretched shape is fixed by cooling the polymer below T_m. When the stretched polymer is heated above T_m, the polymer returns to the original shape. The heat-shrinkable tubes are used for sealing and tightly connecting electrical wires. The film is also used for compact lapping.

9.3 Photo-responsive shape memory effect

When shape memory effect is switched by light, the materials may find various applications.[10] It is known that some molecules are transformed under photo-irradiation to other isomers, which return to the initial state either thermally or photochemically. This reaction is referred to as a photochromic reaction.

$$A \underset{hv'}{\overset{hv}{\rightleftharpoons}} B$$

Table 9.4 shows some typical photochromic reactions, which include (a) *trans–cis* isomerization, (b) zwitter ion formation, (c) ionic dissociation, and (d) ring-formation and ring-cleavage. These reactions are always accompanied by certain changes in the physical and chemical properties, such as dipole moment and geometrical structure. These changes induce shape changes of the polymers, in which the chromophores are incorporated.

Table 9.4. *Photochromic compounds*

Type of reaction	Example
(a) *Trans–cis* isomerization	
(b) Zwitter ion formation	
(c) Ionic dissociation	
(d) Ring-formation and ring-cleavage	

Polymer fabric and film

The most simple example is a polymer having azo-dyes as the crosslinking agents (Figs. 9.7 and 9.8). The following polymer was synthesized and the photostimulated shape change was measured.[11]

R: OC_2H_4O and

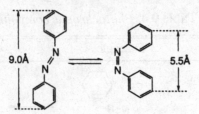

Fig. 9.7. *Trans–cis* photoisomerization of azobenzene.

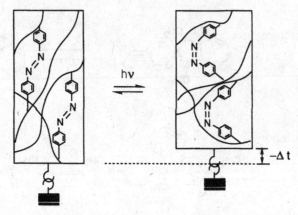

Fig. 9.8. A schematic illustration of a photo-shrinkable polymer film.

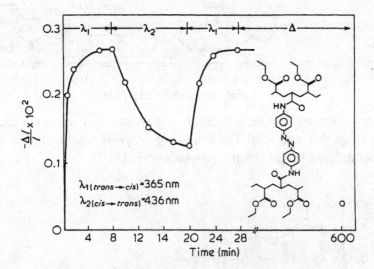

Fig. 9.9. Photostimulated reversible shape changes of poly(ethylacrylate) network with azobenzene crosslinks.[11]

When the azo-dye was converted from the *trans* to the *cis* form by ultraviolet irradiation, the polymer film shrank as much as 0.26%, while it expanded as much as 0.16% by visible irradiation, which isomerizes the azo-dye from the *cis* to the *trans* form. In the dark the film reverted to the initial length, as shown in Fig. 9.9.

Polymer gels

Polyacrylamide gels containing a small amount of triphenylmethane leuco-hydroxide or leucocyanide groups were prepared.[12] The leucohydroxide or leucocyanide groups dissociate into ion pairs by photoirradiation, and recombine with each other in the dark, as shown in Table 9.4.

The gels were swollen to equilibrium by allowing them to stand in water overnight. Then the change in size induced by ultraviolet light was measured.

A disk-shaped gel (10 mm in diameter and 2 mm in thickness) having 3.7 mol% triphenylmethane leucohydroxide residues showed photostimulated dilation in water, as shown in Fig. 9.10. Upon ultraviolet irradiation the gel swells by as much as 3 times its original weight in 1 h. The gel contracts in the dark to its initial weight in 20 h. The cycles of dilation and contraction were repeated several times. When the leucohydroxide residues were replaced with leucocyanide residues, the gel swelled as much as 10 times in weight and 2 times in length.

When one side of a rod-shape leucocyanide gel was irradiated, the gel showed bending as shown in Fig. 9.11. In the dark the gel again returns to the initial straight shape. When two spots of the gel were irradiated, the gel bent as shown in Fig. 9.11(c). The gel shape can be controlled at will by switching the illumination light on or off.

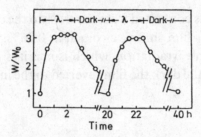

Fig. 9.10. Photostimulated reversible shape changes of a polyacrylamide gel having pendant triphenylmethaneleucohydroxide groups in water at 25°C. $\lambda = 270$ nm.[12]

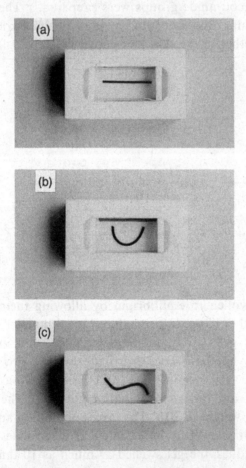

Fig. 9.11. Photostimulated shape changes of a rod-shape polyacrylamide gel having pendant triphenylmethane leucohydrocyanide groups in water at 25°C: (a) before photoirradiation; (b) one side of the gel was irradiated; (c) two spots of the gel were irradiated.

Table 9.5. *pH Sensitive polymers*

$$\left(CH_2-\underset{R}{CH}\right)_{1-x}\left(CH_2-\underset{\underset{O^-}{\overset{|}{C}=O}}{CH}\right)_x \underset{-H^+}{\overset{H^+}{\rightleftharpoons}} \left(CH_2-\underset{R}{CH}\right)_{1-x}\left(CH_2-\underset{\underset{OH}{\overset{|}{C}=O}}{CH}\right)_x$$

$$\left(CH_2-\underset{R}{CH}\right)_{1-x}\left(CH_2-CH\right)_x \underset{-H^+}{\overset{H^+}{\rightleftharpoons}} \left(CH_2-\underset{R}{CH}\right)_{1-x}\left(CH_2-CH\right)_x$$

(aromatic ring with O⁻) → (aromatic ring with OH)

$$\left(CH_2-\underset{R}{CH}\right)_{1-x}\left(CH_2-CH\right)_x \underset{H^+}{\overset{-H^+}{\rightleftharpoons}} \left(CH_2-\underset{R}{CH}\right)_{1-x}\left(CH_2-CH\right)_x$$

(pyridinium $\overset{+}{N}H$) → (pyridine N)

$$\left(CH_2-\underset{R}{CH}\right)_{1-x}\left(CH_2-CH\right)_x \underset{H^+}{\overset{-H^+}{\rightleftharpoons}} \left(CH_2-\underset{R}{CH}\right)_{1-x}\left(CH_2-CH\right)_x$$

(aromatic ring with $H\overset{+}{N}(CH_3)(CH_3)$) → (aromatic ring with $N(CH_3)(CH_3)$)

The gel expansion is interpreted as follows. Upon photoirradiation the leuco-derivatives in the polymer chains dissociate into ions. The formation of ions, fixed cations and free anions generates an osmotic pressure difference between the gel and outer solution, and this osmotic effect is considered to be responsible for the gel expansion.

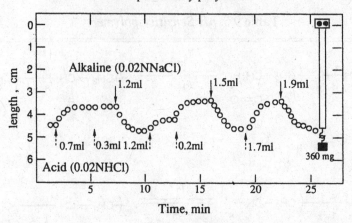

Fig. 9.12. Shape changes of a poly(acrylic acid) film by alternate addition of acid (0.02 N HCl) and alkali (0.02 N NaCl) in water. Initial length: 5 cm; weight: 350 mg.[14]

9.4 Chemo-responsive shape memory effect

The shape of polymers can also be changed by chemicals, when the polymers contain reactive pendant groups.[13] The most extensively studied are polyelectrolytes. Table 9.5 includes several polymers, which dissociate into ions by changing pH. When these polymers are connected to each other by crosslinking agents, the polymer networks make a gel or a film, which changes shape on the addition of acid or alkali in water. Figure 9.12 shows the reversible shape change of a poly(acrylic acid) film by alternate addition of acid and alkali.[14] The film (5 cm in length with a load of 3.5 N) shrinks as much as 1 cm by the addition of 0.02 N HCl and expands again to the initial shape by the addition of 0.02 N NaCl. The cycles can be repeated as many as 2000 times.

If we subject the gels to mechanical work, the contraction force should be large. In order to achieve the strong contraction force of the gel, the gels must be constructed with a large number of crosslinked networks or with a rigid chemical structure. One example of such gels is a PAN gel.[15]

The gel is prepared by peroxidation and subsequent hydrolysis of poly-acrylonitrile fiber. The gel showed large swelling in alkaline solution and collapsed in an acidic solution at $750 \, g/cm^2$ load. The swelling change in length was about 80% and the contraction/elongation response time was less than 20 s. The maximum contraction force was $ca \, 12 \, kgf/cm^2$. The contraction force is comparable to that of living skeletal muscle. The contraction force in the gel fiber is dependent on the aromatization degree of the polymer. When the degree was high, the contraction force increased. The rigid chemical structure is necessary to achieve a strong contraction force.

References

1. A. Makinouchi, *Sosei to Kako,* **20** (1976) 618.
2. S. Takei, in *Development and Applications of Shape Memory Polymers* (in Japanese), ed. M. Irie (CMC, Tokyo, 1989), p. 11.
3. Japan Kokai, JP 59-53528 (Nippon Zeon Co. Ltd).
4. M. Ishii, in Ref. [2], p. 25.
5. Japan Kokai, JP 62-192440 (Kuraray Co. Ltd).
6. M. Karouji, in Ref. [2], p. 44.
7. Japan Kokai, JP 63-179955 (Ashan Chem. Co. Ltd).
8. H. Tobushi, S. Hagashi and P. H. Lin, *Nippon Kikai-Gakkai Ronbunshu,* **60** (1994) 1676.
9. K. Ueno, in Ref. [2], pp. 58, 127.
10. M. Irie, *Adv. Polym. Sci.,* **94** (1990) 27.
11. C. D. Eisenbach, *Polymer,* **21** (1982) 1175.
12. M. Irie, D. Kungwatchakun, *Macromolecules,* **19** (1986) 2476.
13. M. Irie, *Adv. Polym. Sci.,* **110** (1993) 49.
14. W. Kuhn, A. Katchalsky, H. Eisenberg, *Nature,* **165** (1950) 514.
15. N. Okui and S. Umemoto, in *New Functionality Materials,* ed. T. Tsuruta, M. Doyama, M. Seno, Y. Imanishi (Elsevier, Amsterdam, 1993) p. 165.

10

General applications of SMA's and smart materials

K. N. MELTON

10.1 Introduction

Shape memory alloys (SMA's) have many unique properties which lend themselves to a variety of applications. Large heat recoverable strains of up to 8% are obtained in NiTi alloys over a relatively narrow temperature range. Although these strains are small compared to shape memory polymers (see Chapter 9), if a NiTi alloy is constrained to physically prevent the SME from occurring, then stresses of up to 700 MPa can be generated, which are enormous compared to shape memory polymers.

Between these two extremes of 8% strain with little or no force generation, and 700 MPa stress with little or no recoverable strain, lies a huge opportunity to design components with different strain outputs, different amounts of external work deliverable on heating, and a range of cycle lives. Inevitably there are trade offs – it is unfortunately not possible to create components capable of recovering 8% against a force of 700 MPa and repeating this for many millions of thermal cycles. Nevertheless, an enormous variety of applications has been suggested using this range of deliverable properties available to the designer. It is estimated that there are over 10 000 patents or utility models based on SMA's, however, not all of these 10 000 patents have led to successful commercial uses of SMA's. In addition to the normal shape memory effect, where strain is recovered on heating, the superelastic effect can also be used where the material goes round a transformation cycle with no external temperature change, simply by loading and unloading just above A_f. Many recent applications use superelasticity.

In the next three chapters, the applications of SMA's will be described in detail. The most prolific application type, i.e. where most of the 10 000 plus patents have been filed, is as actuators and this is the subject of Chapter 11. Much of the recent focus on SMA commercialization has been on medical and

220

dental applications, particularly those using superelasticity, and these applications will be discussed in Chapter 12. The remainder of the present chapter will be devoted to general applications of SMA's, in other words all applications excluding actuators and medical. The format used will be to describe the history of the development of the application, to discuss the choice of particular SMA, and then to describe current trends in these applications.

Earlier chapters have covered shape memory ferroelectrics, and shape memory polymers. These kinds of materials have recently been described by the term 'smart materials,' and a brief overview of their applications will also be given.

10.2 History of applications of SMA

The now well-documented first large scale application of SMA's in 1971 was as a coupling to connect titanium hydraulic tubing in the Grumman F-14 aircraft. This was a classic example of technology being adapted to a real market need. The decision to use titanium tubing for weight savings was already made, but difficulty was experienced in providing a reliable coupling system. Raychem already sold heat shrinkable polymeric tubing known as Thermofit® tubing, hence NiTi couplings which shrunk on removal from liquid nitrogen were created and known as Cryofit® couplings. SMA's were now on the map. However, after the initial success in meeting a real market need, the next 15 years or so were spent with a technology in search of a market, a classic technology push scenario.

With Raychem as the leader, the niches found for this technology were in the U.S. defense world, where higher prices for superior performance were accepted, and electrical connectors and fasteners resulted. As Japanese companies were successful in SMA technology, particularly after a MITI initiative in the early 1980's, commercial applications in products such as air conditioning vents[1] and bra wires[2] were found. These were mostly market-pull, where the breadth of engineering knowledge of SMA's in Japan helped make this happen; the SMA understanding was not limited to a few individuals in the alloy and supplier base, it was spread throughout the potential user base. Meanwhile, starting at a low level in the early 1980's, orthodontic and then other medical applications, mostly of superelasticity but occasionally of the SMA effect, started to grow because of the excellent fit between the requirements of the marketplace and what the technology provides. Medical applications probably account for the largest dollar value of SMA today, although couplings are probably still ahead on a tonnage basis, but losing fast.

10.3 SMA couplings

10.3.1 Principles

The couplings are made by machining a cylinder, usually with circumferential sealing lands on the inner diameter (ID), while the alloy is in the austenitic condition, cooling to form martensite or at least to where martensite can be stress induced, and expanding by forcing a tapered mandrel through. If tubing is inserted whose outer diameter (OD) is intermediate between the as-machined and as-expanded ID of the coupling, then when the coupling is heated through its A_s–A_f transition it will recover, compress the tube and create an excellent joint. A recent overview of couplings is given in Ref. [3].

NiTi-SMA couplings provide a highly reliable joint for demanding applications such as aircraft hydraulic couplings, but their relatively high cost has been a barrier to more widespread commercial use. Cu-based and more recently Fe-based couplings are cheaper, but because of the lower recovery force of these alloys, they result in a lower performance system.

10.3.2 Alloys

(a) NiTi-based

NiTiFe was the original alloy used for the F-14 couplings. Because the operating temperature range goes down to − 55°C, the only way, in 1971, to prevent the coupling transformating to martensite (and thus softening and leaking) during service, was to use an alloy with a very cold M_s. Binary NiTi alloys with low M_s are Ni-rich and unstable in the sense that stress relief heat treatments after machining can cause M_s shifts. Fe is used to substitute for Ni and depress M_s, but remaining in the region of M_s versus Ti content which is relatively flat. In other words, a material with a relatively wide melting and processing window results. The expansion was done in liquid nitrogen and recovery took place on warming up to room temperature. The only disadvantage of this is that the product must be shipped in liquid nitrogen. The NiTiNb wide hysteresis alloy was the result of a focused research effort[4] at supplying this market need, a NiTi based coupling which could be shipped at ambient temperatures in the expanded condition. Liquid nitrogen dewars were no longer required. Whereas with NiTiFe, mother nature was the installation tool, for NiTiNb some form of heater is required. One of the advantages of NiTi aircraft hydraulic couplings is that they can be installed in difficult to access areas, and the tubing can be placed close together or close to a bulkhead, etc. Requiring the use of a heater takes away this advantage. So far, NiTiNb couplings have only really proven themselves as repair couplings. If cost competitive couplings can be made, however, then the NiTiNb alloy seems to offer the best combination of performance and ease of general installation, and would clearly be the alloy of choice for industrial couplings.

(b) Cu-based

Cu-based SMA's can be thermally preconditioned such that their A_s temperature is raised for the first heating cycle,[5] thus again allowing couplings to be shipped at room temperature. However, the mechanical force which such couplings or fasteners can exert, and also the amount of recoverable strain are lower than for NiTi alloys. This is balanced by their lower material cost. The use of rings of a Cu-based alloy to attach braid to the back of adapters was a successful application for such alloys,[6] but these rings were later replaced by welded wire rings of NiTiNb[7] which eliminated the need for the machining operation. Cu-based components thus had some success in the U.S., primarily in defense type applications, but only limited acceptance in commercial markets mainly because of their overall cost including machining, etc. It is possible that profitable markets for Cu-based couplings could be found if they were manufactured at low cost, but their extensive deployment on a wide range of pipes and tubes is unlikely, mainly because of the lower recoverable strain and the tolerance limitations this imposes on the dimensions of the components to be joined.

(c) Fe-based

Shape memory effects have been observed in a range of Fe-based alloys, but those based on FeMnSi have received most attention for coupling applications. Good SME was reported[8] in polycrystalline alloys, and corrosion resistance was improved by adding Cr and Ni.[9] However, in a recent paper[10] maximum recoverable strains in the region of only 2 to 3% were reported. These FeMnSi based alloys suffer from the further disadvantage of having the force generated during constrained recovery being very sensitive to the amount of unresolved strain.[10] This is in marked contrast to NiTi based alloys where the force is only weakly dependent on the amount of constraint over a wide range of strain.[11] To ensure an adequate joint at low unresolved recovery, i.e. where the clearance between the coupler and pipe is large, a screw-type design was developed.[10] The ends of the pipes to be joined were threaded, as was the inside of the coupling after its expansion. After screwing the components together, the coupling is heated and the SME then gives a significant improvement in clamping force. However, since this requires cutting a thread in the pipe, it can never be used for specifications requiring the pressure

Table 10.1. *A qualitative comparison of the properties of couplings made from the three types of alloy systems*

	NiTi-based	Cu-based	Fe-based
Recoverable strain	high	low	low
Force generated	high	low–medium	low
Corrosion resistance	high	low	medium–high
Cost	high	low	low

performance of the coupling to equal or exceed that of the rating of the pipe. It is not yet clear how successful these Fe-based couplings will be commercially, but their lower cost compared to NiTi might offer some advantages, particularly for large pipe sizes. Table 10.1 compares qualitatively the performance of couplings from NiTi-based, Cu-based and Fe-based alloys.

10.3.3 *Advantages of SMA couplings*

The classical advantages of the original cryogenic NiTi couplings for aircraft hydraulic tubings are:

- Lightweight. This is partly a consequence of the lower density of titanium, but also, compared to swaged fittings there is no spring back once the swaging tool is removed. This allows the use of a compact design. Another factor helping to reduce weight is the use of the specially processed alloy to significantly increase its recovery strength,[12] such that a thinner walled coupling can achieve the same result.
- Easy, craft-insensitive installation even in difficult to access areas. If the hydraulic system is designed to use SMA's, the tubing can be packed closer together.
- Proven reliability. SMA couplings have an excellent track record.

10.3.4 *Trends in SMA couplings*

With the recent downturn in the defense industry, a logical move for NiTi SMA couplings would be into commercial aircraft. However, historically, commercial airplanes have used dematable couplings. This led to the development of Cryolive® fittings where the SME is used to install a flareless end fitting, Fig. 10.1. The Cryolive assembly as-shipped includes a plastic cap which acts as an end stop during installation, but which also helps to thermally insulate and allows a comfortable installation time. The tube is inserted into the assembly, the SMA sleeve allowed to recover, then the plastic cap is removed and the nut screwed onto the connection piece. The SMA sleeve is

Fig. 10.1. Cryolive® couplings: (top) the shipping assembly with nut and plastic cover; (middle) showing the Cryolive SMA component shrunk on the tube with the nut pulled back; (bottom) the nut is threaded onto the mating component. (Photographs courtesy of AMCI, Menlo Park, Carlifornia)

contoured such that it forms a compression sleeve when the parts are assembled.

With their present cost, it is likely that NiTiNb alloys will mainly be used for specialty applications such as battle damage repair, or nuclear repairs. A possible approach to get the necessary low costs might be for a company already in the industrial coupling market to adopt the NiTi technology. Lower cost Fe-based alloys seem to have some potential, but probably only in low performance applications where the system pressure rating is less than the capability of the pipe or tube. Because of corrosion compatibility, use of these couplings might be restricted to steel pipe, but that is still the most widely used class of piping material.

An interesting recent development is the use of SMA not as a coupling to replace a weld, but as a reinforcement of a pipe weld. Prestretched wires of NiTiNb are wrapped around a fitting installed over a pipe weld, Fig. 10.2. On heating, the wire contracts and compresses the fixture. This technique can be used either to reinforce the pipe weld, in other words to take the full pipe load in case of rupture, or to create compressive stresses to prevent cracking in the weld region. The use of wires enables a wraparound configuration which can be installed after welding.

Fig. 10.2. Strength reinforcement of a pipe weld using SMA wire wraps. (Photograph courtesy of ABB Atom, Sweden)

10.4 Electrical connectors

The second major application of SMA's to be developed was their use in electrical connectors. Unlike most tube or pipe couplings, electrical connectors must be able to sustain several connect/disconnect cycles. This is done by utilizing the strength differential between the austenitic and martensitic phases, usually in combination with a compliance such as a mechanical spring.[13] In the original Cryocon connector, Fig. 10.3, the slit ends of a Cu–Be tube are formed open to act as a bias spring. A martensitic SMA ring is installed, such that on heating and recovery the ring compresses the spring. This is illustrated schematically in Fig. 10.4(a), where mechanical equilibrium is reached at point A in the austenitic condition and point B in the martensitic. On thermal cycling, the displacement difference B–A is then obtained. If in the open condition B a suitably sized connector pin is inserted, then on heating the connector will close. In fact, once contact is made with the pin, point C in Fig. 10.4(b), then further recovery of the SMA ring will elastically compress the pin and the force will increase more rapidly. Care must be taken that the stress at point A does not increase to the extent that the austenitic yield strength is approached, since this will result in poor cyclic performance.

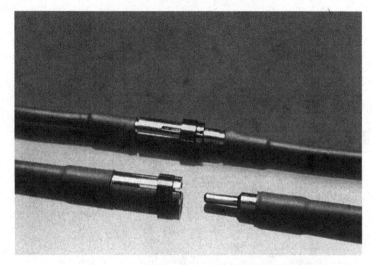

Fig. 10.3. Cryocon® electrical connectors, mated (top) and demated (bottom) showing the slit socket, NiTi ring and pin.

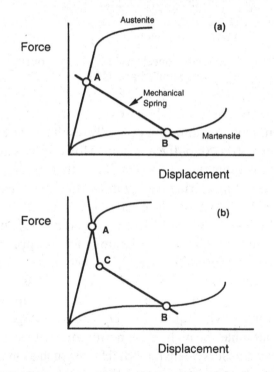

Fig. 10.4. A schematic illustration of the force–displacement relationship between an SMA element and a mechanical spring. In (a) the system will displace between points A and B on thermal cycling. In (b) on heating, contact is made with the pin at point C, the line CA then represents the stiffer compliance of the pin.

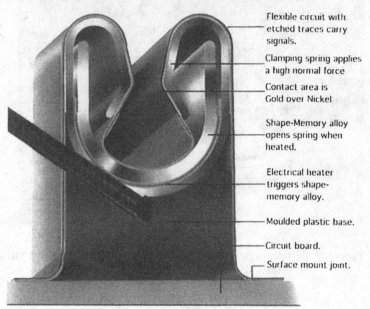

Flexible circuit with etched traces carry signals.

Clamping spring applies a high normal force

Contact area is Gold over Nickel

Shape-Memory alloy opens spring when heated.

Electrical heater triggers shape-memory alloy.

Moulded plastic base.

Circuit board.

Surface mount joint.

Connector cross section.

Fig. 10.5. The Betaflex™ electrical connector. (Photograph courtesy of Beta Phase Inc., Menlo Park, CA)

The connector decribed here is for the case that the connector is closed in the austenitic condition. This configuration generally results in the highest contact force, using the higher strength of the austenite. However it is also possible to use the SMA/spring combination in reverse, such that on heating, the SMA displaces against the spring storing energy in it. On subsequent cooling, this stored energy is used both to make the electrical contact and to re-deform the SMA component. In this case the SMA has an M_s temperature typically above room temperature, and installation can be done for example by electrically heating the SMA. An example of this kind of connector is shown in Fig. 10.5.

For the Cryocon connector, the SMA ring typically has an M_s (at load!) below the lowest operating temperature of the connectors, opening is done by cooling the connector, e.g. with liquid nitrogen. The advantage of the Cryocon type connector is that it has the highest connector forces. This is important for connectors which need to survive high inertial forces such as in missile control systems. Shock and vibration resistance of these SMA connectors is the best available. In comparison, the high M_s type connector offers the convenience of ease of installation at the expense of lower contact forces, but for non-defense applications the connectors can be designed such that they are still more than adequate.

10.4.1 SMA selection

A common feature of both of these types of connector is that the alloy should have the greatest strength differential between the austenite and martensite phases. A soft martensite allows point B in Fig. 10.4 to be as far as possible to the right, and particularly for the Cryocon type connector, a strong austenite prevents point A (Fig. 10.4(b)) from causing yielding. Alloys based on NiTiCu are typically used to meet these criteria.

10.4.2 Trends in SMA connector applications

As noted above, Cryocon was originally developed as a defense type product, providing superior performance. However, because of the small absolute displacements available on thermal cycling, manufacturing tolerances were extremely tight. Furthermore, these tiny rings were individually machined. Consequently the connectors were correspondingly expensive.

The next generation, Fig. 10.6, still used individually socketed pins, but the SMA drivers were now stamped reducing the cost per line. More importantly, the design was changed such that the tolerances could be opened to allow the use of industry standard pins. The Cryocon connectors only worked by tightly controlling the tolerances of *both* parts of the connector. Nevertheless, the cost/performance envelope of this new connector did not allow widespread commercial use.

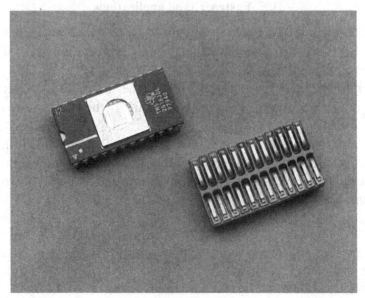

Fig. 10.6. Top view of the Crotact™ electrical connector.

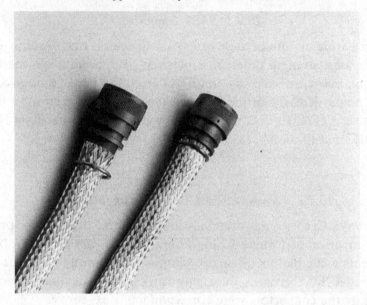

Fig. 10.7. A welded ring of NiTiNb is used to fasten shielding braid to the back of a harness connector.

The connector shown in Fig. 10.6 has one SMA driver making many line connections. Designs such as these are opening up commercial markets for SMA connectors.

10.5 Fastener type applications

The term 'fastener' is used here to mean an SMA component which is used to fasten or join together things other than signal carrying electrical connectors or tube or pipe couplings.

An example of this type of application is the use of SMA rings to join screening braid to adapters, see Fig. 10.7. Originally these rings were machined, in the case of NiTi from cryogenic alloys, a combination which made them expensive and somewhat difficult to install. The advent of welded rings from NiTiNb alloys has changed this considerably, although for some applications the weld-bead is unacceptable, and for applications needing high stress, machined rings are still used.

Advantages of SMA fasteners are:

• Can be installed to precisely locate components with a controlled pre-load, e.g. bearings on a shaft, Fig. 10.8.
• Good shock, vibration and thermal cycling properties.
• Can provide a hermetic seal, Fig. 10.9.

Fig. 10.8. Prototype of a gear blank assembled on a shaft using a NiTiNb ring. The dark patch on the ring is the thermochromic paint indicator that the correct installation temperature has been reached. (Photograph courtesy of Intrinsic Devices, San Francisco, CA)

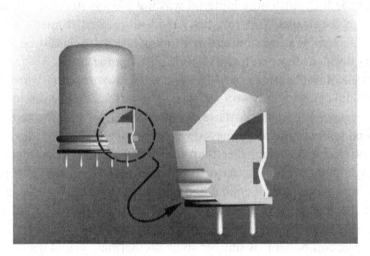

Fig. 10.9. Schematic cut-away of a hermetic seal made using a NiTi ring. (Photograph courtesy of Intrinsic Devices, San Francisco, CA)

- Easy to install, performance is relatively operator friendly. With NiTiNb rings the installation can be done electrically and could be automated.

Despite the performance advantages of these fastener rings, the costs have still been too high to allow their widespread use. It remains to be seen whether generic (i.e. non-custom) fastener rings can be made cheaply enough to be widely used. To date, all the fastener successes of SMA have been in higher performance niche markets.

10.6 History of applications of superelasticity

The first commercial use of the highly elastic properties of SMA's was as orthodontic wires for the correction of teeth. In fact however, the material originally used was one which was martensitic at ambient temperature, use being made of the elastic properties of cold-worked martensite. Currently the majority of such arch wires use the stress-induced martensite, providing many advantages over conventional spring materials. The next application was also a medical one, the Mammelok® breast hook used to locate and mark breast tumors so that subsequent surgery is more precise, requiring less radical tissue removal.

Many of the possible applications of SMA being pursued by medical practitioners in the mid 1980's were attempting to use the thermally activated memory effect. However, temperature excursions tolerated by the human body are very limited, requiring very precise control of the A_s temperature. Also, with only a narrow ΔT available as a consequence of this, the maximum strength of the fully austenitic material was not always realized, giving additional design constrains. Superelasticity is ideally suited to medical applications since the temperature window of optimum effect can readily be located to encompass ambient temperature through body temperature.

Since medical applications of SMA's including superelasticity are the subject of the final chapter, only non-medical applications will be discussed here. The first widespread non-medical use of superelasticity was as the support wire in a bra, and many of the recent new applications of NiTi based alloys use the superelastic characteristics rather than the shape memory effect itself. When considering shape memory actuator type applications, other competing solutions such as wax motors or bimetals must be reviewed since they are often more cost effective. However, the superelastic property is unique; superelastic NiTi is simply the springiest metal readily available. Furthermore, the design of superelastic components is simpler since no thermal cycle is needed and no mechanism required to reset the device on cooling. A good example of the use of the high springiness of NiTi is as the antenna of portable phones in Japan.

From a materials supplier point of view, superelastic applications are easier to support, since in most cases the superelastic effect is used in the temperature range from ambient to body temperature. This allows most uses to be covered by a single alloy, whereas the requirements for thermally activated devices are broad, necessitating many alloys with different M_s and A_s temperatures. The slightly Ni-rich binary alloy is thus the most commonly used for superelastic applications. Where a narrow stress hysteresis is required, in other words where the restoring or unloading force should be as close as possible to the

Fig. 10.10. A reinforcing wire of superelastic NiTi is used to retain the shape and comfort of the heel of a shoe. (Photograph courtesy of Furukawa Electric)

initial deforming force, then NiTiCu based alloys with their narrower hysteresis can offer some advantages.

Following on from the bra-wire application, a recent development is to use a superelastic wire in the heel of a shoe, Fig. 10.10. The wire retains the shape of the shoe and improves the fit and comfort over time.

Another recent application is as the headband of headphones, Fig. 10.11, used by Sony for their MiniDisc Walkman® in Japan. Use of superelasticity allows the headphones to be folded up into a compact egg shape for carrying (hence the name 'Eggo'), and also gives a constant low pressure on the ears when in use. The headphones are thus very comfortable to wear, which I can confirm having worn a pair for most of the ten hour flight back from Japan!

The same features of light weight, retention of shape even if misused and being more comfortable to wear because of the steady pressure were key for another application of superelastic elements in eyeglass frames, Fig. 10.12. NiTi elements are used in the nose and brow bridges and as the temple pieces. Critical to the success of this product range was the development of joining

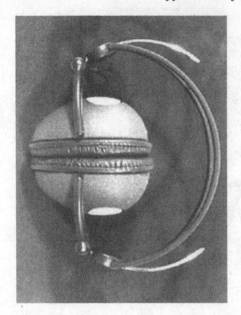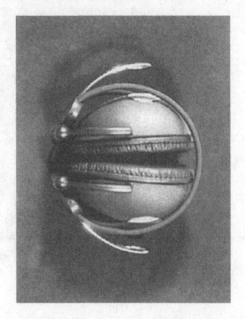

Fig. 10.11. Headphone for Sony's 2.5 inch MiniDisc Walkman: left, unfolded; right, folded. Two pieces of 1.5 mm superelastic wire are used to make a headband. (Photograph courtesy of Kato Hatsujo Kaisha Ltd., Japan)

techniques to enable the assembly of the components and plating to provide the attractive finish required.

A common theme of all these successful non-medical applications is in fact their association with the human body, which puts a convenient limit on the temperature range any superelastic component needs to survive in service. This is also one of the success factors for medical applications of superelasticity.

10.7 Selection criteria for SMA applications

Most people working on industrial applications of SMA's receive enthusiastic telephone calls from engineers who have just read an interesting article on SMA's and would like to use them. Although purely subjective, the following criteria are offered to screen potential applications.

1 Cost

SMA's are becoming more widely used and consequently their costs are decreasing. Nevertheless, they are still relatively expensive on a unit weight basis, and particularly when an SMA solution is one of several being evaluated, a rapid cost estimate will indicate whether further study is warranted.

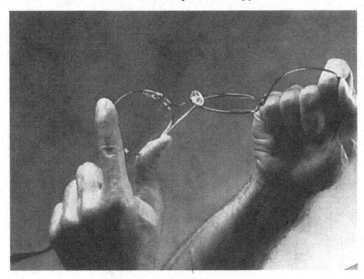

Fig. 10.12. Superelastic NiTi eyeglass frames. (Photograph courtesy of NDC, Fremont, CA)

2 Does SMA provide a unique solution?

In many ways this relates to item 1. If there is an alternative way of delivering the same function (such as a bimetal versus SMA actuator) then SMA will often lose on cost grounds. Many prospective customers are surprised to be given an alternative solution after calling for SMA information! However if there are limitations to the competitive solution, for example space requirements or size, then SMA can be preferred. An example of this is the anti-scald shower device.

3 Alloy requirements. Can the potential application be met with an existing alloy?

To some extent this is volume related. Any new applications requiring an alloy with a different M_s for example, is clearly of interest if thousands or (better!) millions of components are anticipated. But melting and qualifying a new alloy for a few hundred springs makes no economic sense. I once estimated that the cost of developing and characterizing a high temperature PtNiTi or PdNiTi alloy far exceeded the total expected profit from sales of the high temperature actuator, even though the application would support the material cost of the alloy once developed.

This is one big advantage of superelasticity as noted above – a very limited number of standard alloys will cover most applications.

Table 10.2. *Examples of smart materials*

Input or stimulus	Output or response	Material	Application	Reference
1. overcurrent (or temperature increase)	resistance increase	ceramic e.g. La doped $BaTiO_3$, $Pb(ZrTi)O_3$ polymer e.g. C-black filled polyethylene	thermistor	14
			overcurrent protectors	15
2. overvoltage	resistance decrease	varistor e.g. Bi doped ZnO	overvoltage surge protector	16
3. change in oxygen partial pressure	electric signal	Y_2O_3 doped ZrO_2	oxygen sensor, the electrical signal can be used as part of a feedback mechanism e.g. for vehicle exhaust	14
4. deformation or strain electric signal	electric signal deformation or strain	piezoelectric material ceramic e.g. PZT polymer e.g. PVDF	active noise control devices pressure and vibration sensitizing (intruder alarm) vibration control (road roughness ride)	17, 18 21, 22
			acoustic emitters	23 24
5. pH change	swelling or contracting	polymeric gel	artificial muscle	25
6. electric signal	viscosity change (increase with electric field on)	electro-rheological fluid (e.g. 35% cornstarch 65% silicone oil)	torsional steering system damper	20
7. temperature	electric signal	pyroelectric material (ferro-electric) e.g. Pb(Zr, Ti)O_3	personnel sensor (open supermarket door)	24
8. humidity change	capacity change	polymer e.g. thin film cellulose esters ceramic e.g. Al_2O_3	humidity sensors	26
	resistance change	polyelectrolyte e.g. poly(styrene) sulfonate ceramic e.g. $MgCr_2O_4$–TiO_2		

4 Technical requirements

Having gone through these filters, will the alloy's performance meet the requirements of the applications? Many customers would like 'impossible' combinations, such as 8% heat recoverable strain, high work output and ten million cycles.

One important factor to consider in all potential applications is how to manage the interface between the material supplier and the end user. We are still a long way away from selling NiTi with a data sheet and alloy sample (although we're coming close for superelasticity). The material specialist and design engineer need to work closely together, this has been the case in all successful products at Raychem.

10.8 Smart materials

The term 'smart materials'[14] has been most often used in the context of smart or adaptive structures, with the emphasis being on control of the stiffness or shape of an aerospace platform. In many of these studies the sensor and actuator functions were separate, with an electronic control circuit to provide the necessary feedback. Although it is often used to imply the sensor capability only, for the purpose of the present work, a smart material is defined as one which can combine both the sensor and actuator functions. In other words, a material which changes one of its property coefficients in response to an external stimulus, and where this change in coefficient can be used to control the stimulus. This definition is then consistent with the behavior of shape memory alloys when used as actuators, for example the well-known use as a control for air conditioning systems (1) where the SMA element senses a change in temperature, responds by changing its length and this length change activates a valve to direct air flow. Smart materials include ceramics, polymers and metals, and the nature of the external stimulus and the corresponding property change can vary widely. Table 10.2 summarizes some of the possibilities. If a linear relationship is required between the stimulus and the response, then some form of controlled feedback might be needed through some kind of signal conditioner, microelectronics being of sufficiently low cost nowadays. However, in many cases, a step function is adequate. For example, carbon black filled polyethylene can show a resistance increase with temperature of several orders of magnitude.[15] The effect is nonlinear, but is so large that it can be used to protect against overcurrent. Such an element responds to a current increase by increasing its temperature, at a critical value of which its resistance increases dramatically, reducing the current flowing through it to effectively zero. Another similarity between SMA and many smart materials as defined here, is that their 'adaptiveness' can be programmed. Thus for SMA's it is done by the shape setting process during annealing whereas for piezoelectric actuators it is done by alignment in an electric field.

Compared to shape memory alloys, arguably the first smart material, many others are in their infancy with respect to developing applications. In the opinion of the present author, there is a huge potential for ceramic smart materials, since they usually have an electrical charge as either the input or the output. As we all know, many everyday items now have some form of electronics in them. It is the ability of electroceramics to be directly integrated into electronics or electrical devices which will be a significant factor in driving their growth in applications.

A major development will probably be multilayer materials. One advantage of these is to increase the response, either to provide a much larger output signal for the same stimulus, or to generate the same output from a much smaller stimulus. Ceramic piezoelectric devices have relatively small field induced strains (typically 0.01–0.1% strain at 1 MV/m applied field). A 1 cm tall block of monolithic ceramic could produce 10 μm displacement, but would take about 10 kV. Building up a multilayer device of 0.1 mm layers connected in series mechanically, but in parallel electrically, produces similar displacements, but drive voltages are reduced by two orders of magnitude.

Multilayer constructions might allow several active functions to be combined in the same package by building composites with different ceramics. Thin film techniques might also allow the integration of the control function with the sensor actuator, producing an integrated device, or the integration of the signal conditioner with the sensor such that the output is linear and proportional.

Shape memory alloys were originally viewed as an interesting metallurgical curiosity. Perhaps their rightful place is as a smart material which has found a broad range of applications.

Thermofit® is a registered trademark of Raychem Corporation
Cryofit® and Cryolive® are registered trademarks of AMCI, Menlo Park, CA
Mammelock® is a registered trademark of Mitek Inc.
Walkman® is a registered trade name of Sony.

References

1. T. Todoroki and H. Tamura, *Trans. Japan Inst. Metals*, **28** (2), (1987) 83.
2. H. Horikawa, *Symposium on SMA Applications*, Tokyo, Japan (July 3, 1992) p. 23.

3. M. Kapgan and K. N. Melton, in *Engineering Aspects of Shape Memory Alloys*, ed. T. W. Duerig, K. N. Melton, D. Stöckel and C. M. Wayman, Butterworth–Heinemann (1990) pp. 137–148.
4. K. N. Melton, J. Simpson and T. W. Duerig, *Proc. ICOMAT 1986*, pp. 1053–1058, The Japan Institute of Metals.
5. G. B. Brook, P. L. Brooks and R. F. Iles, U.S. Patent Numbers 4,036,669; 4,067,752; 4,095,999.
6. K. N. Melton and T. W. Duerig, *Metallurgia*, **52** (1985) 318.
7. K. N. Melton, *Professional Engineering* (June, 1991) p. 48.
8. M. Murakami, H. Otsuka, H. G. Suzuki and S. Matsuda, *ICOMAT* (1986) p. 985.
9. H. Otsuka, H. Yamada, T. Maruyama, H. Tanahashi, S. Matsuda and M. Murakami, *ISIJ Int'l*, **30** (1990) 674.
10. H. Tanahashi, T. Maruyama, and H. Kubo, Advanced Materials '93, V/B: Shape Memory Materials and Hydrides (ed. K. Otsuka and Y. Fukai), *Trans. MRS Jpn.*, **18B** (Elsevier Sci. B. V., 1994) p. 1149.
11. T. W. Duerig and K. N. Melton, *Materials Research Society International Symposium Proceedings, Shape Memory Materials*, ed. K. Otsuka and K. Shimizu (1989) p. 581.
12. J. L. Proft and T. W. Duerig, in *Engineering Aspects of Shape Memory Alloys*, ed. T. W. Duerig, K. N. Melton, D. Stöckel and C. M. Wayman, Butterworth–Heinemann (1990) pp. 115–129.
13. E. Cydzik, in *Engineering Aspects of Shape Memory Alloys*, ed. T. W. Duerig, K. N. Melton, D. Stöckel and C. M. Wayman, Butterworth–Heinemann (1990) pp. 149–157.
14. S. Trolier-McKinstry and R. E. Newnham, *MRS Bulletin*, **18** (4) (1993) 27.
15. F. A. Doljack, *IEEE Trans.*, **CHMT-4** (4) , (1981) 372.
16. L. M. Levinson and H. R. Philipp, *J. Appl. Phys.*, **46** (3), (1975) 1332.
17. Y. Sugawara, K. Onitsuka, S. Yoshikawa, Q. C. Xu, R. E. Newnham and K. Uchino, in *Proc. of the ADPA/AIAA/ASME/SPIE Conference on Active Materials and Adaptive Structures*, Nov. 1991, ed. G. J. Knowles, Inst. of Physics Publishing, Bristol and Philadelphia, p. 143.
18. O. L. Santa Maria, E. M. Thurlow and M. G. Jones, ibid. p. 529.
19. F. Geil and L. Matterson, ibid. p. 135.
20. J. R. Salois, ibid. p. 745.
21. J. L. Fanson and J. C. Chen, *Proc. of NASA/DOD Control Structures Interaction Conf.*, NASA CP-2447 Part II, (1986).
22. E. F. Crawley and J. de Luis, *AIAA Journal*, **25** (101), (1989) 1373.
23. H. Tsuka, J. Nakamo and Y. Y. Yokoya, *IEEE Workshop on Electronic Applications in Transportation* (1990).
24. A. J. Bell and I. M. Reamey, *Materials World*, **2** (3), (1994) 125.
25. I. Z. Steinberg, A. Oplatka and A. Katchalsky, *Nature*, **210** (5036), (1966) 568.
26. B. M. Kulwicki, *J. Am. Ceram. Soc.*, **74** (4), (1991) 697.

11

The design of shape memory alloy actuators and their applications

I. OHKATA AND Y. SUZUKI

Symbols

C spring index ($= D/d$)
d wire diameter
D mean diameter of spring
G shear modulus (τ/γ)
G_H shear modulus at high temperature
G_L shear modulus at low temperature
n number of active coils
P load on spring
S stroke length of spring

γ shear strain
γ_H shear strain at high temperature
γ_L shear strain at low temperature
γ_{max} maximum shear strain
δ deflection of spring
κ stress correction factor
τ shear stress
τ_{max} maximum shear stress

The thermal actuator is a device which is capable of converting thermal energy into mechanical energy, usually generating force and displacement. A shape memory alloy actuator (SMA actuator) is one type of thermal actuator, utilizing the shape memory effect to generate motion and force. Although the pipe coupling described in Chapter 10 can be considered a thermal actuator, for the purposes of this discussion we will describe 'two-way SMA actuators' which have the ability to exhibit reversible motion as the temperature either increases or decreases. There are two design techniques which are exploited to achieve two-way movement: a biasing method and a differential method. The biasing technique offers greater flexibility in design and is the most frequently used for a broad range of devices. The differential method provides more precise motion control and finds application in robotics and similar actuators requiring high accuracy of motion.

11.1 Characteristics of shape memory alloy actuators

In considering the design and application of SMA actuators it is important to understand the difference between these devices and the more common metal thermostatic element or wax actuators. Metal thermostats are fabricated from

240

sandwich sheet produced by bonding two or more metals with different thermal expansion coefficients. When such an element is heated or cooled the metal element bends in a direction normal to the clad plane. Deflection of the thermostat element, which can be in the form of a simple cantilever or a spiral, varies linearly with the temperature change. The magnitude of the deflection is small and the forces generated are modest. Some enhancement of the deflection can be achieved by using alternative geometries such as belleville washers or flat spiral springs, but these actuators are not suited to applications which require a large stroke or motion.

Wax stats, or wax actuators, utilize waxes which have a large volume coefficient of expansion or undergo a liquid–solid phase change with an accompanying large expansion or contraction. Typical applications for wax thermal actuators are in the cooling system of automobiles and in domestic hot water space heating systems. Deflection of a wax actuator can be abrupt at the temperature of the phase change, although this behavior can be modified by using a mixture of waxes with different melting points. Since the wax must be encapsulated in a sealed container, and the wax thermal conductivity is low, the thermal response time for these actuators is relatively slow. This limits wax actuators to applications where response speed is not critical.

The actuating force in an SMA actuator is the result of a dimensional change which results from the solid state transformation in alloys which exhibit the shape memory effect. Since SMA actuators do not require the sliding seals or liquid encapsulation of wax actuators their design and operation is considerably simpler. Further, wax actuators are basically linear motion devices, whereas SMA actuators can be designed to operate in tension, compression or torsion and produce linear or rotary motion, and in sophisticated devices complicated three-dimensional motion can be achieved. Where larger motions are required than can be achieved with a simple wire or rod in tension, the use of helical coils makes possible actuators with large motions capable of large force output. Since the motion of the SMA actuator is due to a phase change within the actuator itself, devices with no sliding interfaces can be designed.

Figures 11.1(a) and (b) compare the temperature deflection characteristics of wax actuators, thermostatic elements and SMA actuators. When compared on a force/weight ratio basis, the SMA actuator is clearly superior, and, in fact, this superiority is also evident when comparing force/volume ratios. The force and motion of an SMA actuator are better than either of the other two devices, and in addition to the ability to design a smaller actuator exhibiting equal force, the speed of response of the SMA device is also considerably faster. A comparison of the weight/force characteristics of SMA actuators with thermostats, wax actuators and solenoids is presented in Fig. 11.2. The clear superior-

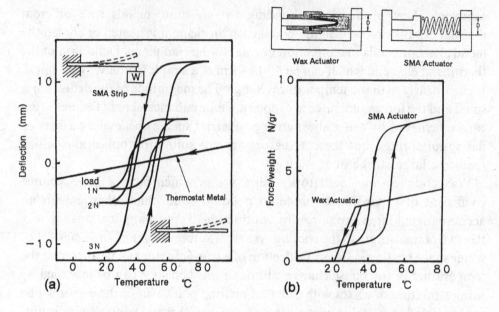

Fig. 11.1. Comparison of the characteristics of thermal actuators. (a) Temperature *vs* deflection curves of a thermostat alloy and a shape memory alloy. (b) Temperature *vs* force/weight ratio curves of a wax actuator and a shape memory actuator.

ity of the SMA actuator is evident, particularly for actuators in the region below 1 g in weight. Even small electric motors cannot provide the performance of SMA actuators where weight is at a premium. The advantages of SMA actuators may be summarized as follows:

(1) Large stroke length/weight ratio
(2) Large force output per unit weight
(3) Flexibility of design of actuator motion direction
(4) Rapid motion at a specified temperature
(5) Insensitive to a wide variety of environmental conditions
(6) Can be designed to be used where cleanliness is critical
(7) Quiet operation.

11.2 The design of shape memory alloy springs

The most commonly used shape memory alloy element in an SMA actuator is a coil spring since this form makes possible a large stroke. As a comparison between the shape recovery of a straight wire and a coil whose closed length is the same, a straight wire of 30 mm length and a 1% strain recovery has a stroke length of only 0.3 mm while an 8 mm diameter helical coil using 30 turns of 1 mm wire will have a stroke of 50 mm. The design of SMA springs

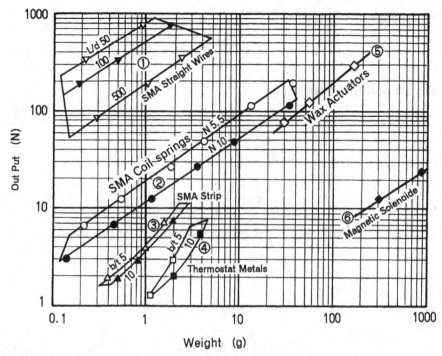

Fig. 11.2. Force output *vs* weight comparison between SMA actuators, wax actuators, thermostat metals and magnetic solenoids: (1) Ti–Ni straight wires – L/d = wire length/diameter; (2) Ti–Ni helical coils – N = active coils; (3) Ti–Ni strips – b/t = strip width/thickness; (4) thermostat metals; (5) wax actuators; (6) magnetic solenoids.

employs the formulae of conventional spring design[1] although there is the complication that the shear modulus in these alloys is not constant. The remarkable feature of shape memory alloys is the dramatic change in modulus as the alloys are heated above and below the phase change transformation temperature. In Ti–Ni shape memory alloys the change in shear modulus from the high temperature to the low temperature phase is approximately 300%. In normal spring design the only requirement is to obtain the relationship between load and deflection, whereas in designing a shape memory alloy spring the change in deflection as a function of temperature must also be calculated. The following are notations and fundamental formulae used in designing helical springs:

D = mean diameter of spring d = wire diameter
n = number of active coils C = spring index
P = load on spring δ = deflection
τ = shear stress γ = shear strain
G = shear modulus κ = stress correction factor.

The equation for deflection

$$\delta = \frac{8PD^3n}{Gd^4}. \tag{11.1}$$

The equation for shear strain

$$\gamma = \frac{\delta d}{\pi n D^2}. \tag{11.2}$$

The equation for shear stress

$$\tau = \frac{8PD\kappa}{\pi d^3}. \tag{11.3}$$

The stress correction factor κ is calculated using Wahl's formula.

$$\kappa = \frac{4C - 1}{4C - 4} + \frac{0.615}{C}. \tag{11.4}$$

The shear modulus G and the spring index C are defined as:

$$G = \tau/\gamma \tag{11.5}$$
$$C = D/d. \tag{11.6}$$

In the design of SMA springs the dimensions are dictated by the force and deflection characteristics required, as well as the space available for the particular device. The dimensions of a spring are the coil diameter D, the wire diameter d and the number of active coils. The force output is defined by the coil diameter and the wire size, while the deflection is defined by the number of active coils. Three parameters, which are characteristics of the particular SMA, are required to calculate the spring dimensions: the maximum shear stress (τ_{max}), the shear modulus at the temperature of the elevated temperature phase (G_H) and the shear modulus of the alloy at the temperature of the low temperature phase (G_L). Typical values for these parameters for commercial Ti–Ni SMA are: $\tau_{max} = 120\,MPa$, $G_L = 8000\,MPa$ and $G_H = 23\,000\,MPa$. SI units are used in these equations except for length which is given in mm. The spring index is an arbitrary parameter and the approximate value of 6 is usually employed since this value is consistent with ease of spring winding. The calculation of a typical spring is given by way of illustrating the procedure.

We will assume a device requiring a force output, the load P, of 10 N and a stroke length of 5 mm.

The parameters of the Ti–Ni SMA are:

$\tau_{max} = 120\,MPa$, $G_L = 8000\,MPa$, $G_H = 23\,000\,MPa$ and a spring index $C = 6$.

The wire diameter is first calculated using Eqs. (11.3) and (11.5):

$$\kappa = (4 \times 6 - 1)/(4 \times 6 - 4) + 0.615/6 = 1.2525$$

$$d^2 = \frac{8PC\kappa}{\pi\tau}$$

$$= 8 \times 10 \times 6 \times 1.2525/3.1416 \times 120 = 1.5947$$

$$d = 1.26 \text{ mm which is rounded to } 1.3 \text{ mm.}$$

The mean coil diameter D is obtained from Eq. (11.6)

$$D = Cd = 6 \times 1.3 = 7.8 \text{ mm.}$$

The next step is the calculation of the number of active coils, n. The stroke length S is equal to the difference between the deflections at high and low temperatures $\delta_L - \delta_H$, and the number of active coils is proportional to the difference between the shear strains at the low and high temperatures. The shear strains are given by Eq. (11.5):

$$\gamma_H = \tau/G_H = 120/23\,000 = 0.0052$$
$$\gamma_L = \tau/G_L = 120/8000 = 0.015$$
$$\gamma_S = \gamma_L - \gamma_H = 0.015 - 0.0052 = 0.0098.$$

Substituting this value of γ_S in Eq. (11.2), we obtain the number of active coils:

$$n = \frac{Sd}{\pi\gamma_S D^2} = 5 \times 1.3/3.1416 \times 0.0098 \times 7.8^2 = 3.47.$$

As we indicated earlier, the maximum shear strain which can be used is dictated by the fatigue life requirements for the actuator, and the low temperature shear strain γ_L at 1.5% is too large for the criterion. In cases such as this the maximum shear strain γ_{max} is employed as a design parameter instead of the value calculated from τ/G_L. The relationship between fatigue life and the design parameters is shown in Table 11.1.[2] In most cases the value for γ_S coincides with the strain at low temperature, γ_L. We can now recalculate the value for n using as the maximum shear strain 1%, and the following parameters:

$$\tau_{max} = 120 \text{ MPa}, \gamma_{max} = 0.01 \ (1.0\%), G_H = 23\,000 \text{ MPa and } C = 6$$
$$\gamma_S = \gamma_{max} - \gamma_H = 0.01 - 0.0052 = 0.0048$$
$$n = \frac{Sd}{\pi\gamma_S D^2} = 5 \times 1.3/3.1416 \times 0.0048 \times 7.8^2 = 7.08.$$

We therefore have the dimensions of the spring which will deliver the required force and deflection and have acceptable fatigue life:

wire diameter 1.3 mm, mean coil diameter 7.8 mm, and 7 active coils.

It should be noted that if the maximum shear strain γ_{max} is specified, the operating strain of the SMA spring must be kept within γ_{max}. For the case of a

Table 11.1. *Design parameters of the Ti–Ni based shape memory alloys*

Alloy type	τ_{max}	γ_{max}	G_H	G_L	Fatigue life
Ti–Ni (R-phase)	120 MPa	1.0%	23 000 MPa	8000 MPa	$> 10^6$
Ti–Ni–Cu	250 MPa	2.0%*	17 000 MPa	2000 MPa	10^4

* 5.0% if used as a straight wire

compression spring, the free or open length is designed so that when the spring is in its closed or coil-bound state the strain is at the designed maximum. When using an extension spring, a position limiter or stop is employed to prevent exceeding the strain limit.

In many cases the maximum spring diameter must be specified because of space limitations imposed by the particular device. For these cases a simple graphical method of defining the other spring dimensions can be employed.[3]

The equations used for this design method are:

$$\kappa C^3 = \frac{\pi D^2 \tau}{8P}, \tag{11.7}$$

$$n = \frac{S}{\pi C D \gamma_S}. \tag{11.8}$$

These equations are derived from Eqs. (11.1), (11.3) and (11.6). The procedure is to first calculate the value κC^3 using Eq. (11.7) and then, using the graph in Fig. 11.3 determine the value for C; knowing C we can calculate the value for d since the dimension C was established. The number of active coils n can then be determined using Eq. (11.8).

The design of a spring with approximately the same characteristics as the previous SMA spring design but using a preselected spring diameter of 8 mm is calculated as follows:

The value for κC^3 is calculated using Eq. (11.7); $3.1416 \times 8^2 \times 120/8 \times 10 = 301$.

Using Fig. 11.3 this value gives a spring index C of 6.1, and from this we determine the wire diameter to be 8/6.1 or 1.31 mm, very close to the previous design.

The active coils $n = 5.0/3.1416 \times 6.1 \times 8 \times 0.0048$ which gives $n = 6.8$.

We have used 23 000 MPa as the value for the shear modulus at high temperature, and Ti–Ni wire fabricated by accepted processing will have this value. Deviations from this value can be expected for different alloy compositions and fabrication processes. When a precise specification of the shear modulus is required for the design of a critical component the value of the

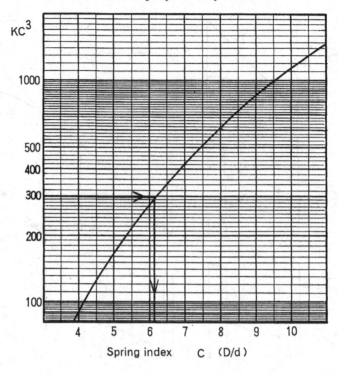

Fig. 11.3. Design chart for a Ti–Ni shape memory alloy spring of round wire. The spring index C is obtained from the calculated value of κC^3.

shear modulus can be experimentally determined using any one of three methods. Since the calculation of G from Eq. (11.5) requires values for τ and δ, each method involves the determination of these values by measuring the force and deflection characteristics of an SMA spring, that is, P and δ. Figure 11.4 illustrates the three methods: (A) the determination of the load–deflection curve at a constant temperature, (B) the determination of the deflection–temperature curve under constant load, and (C) the determination of the load–temperature curve under constant deflection. In designs requiring a precise value for the deflection, method (B) or (C) is recommended.

11.3 The design of two-way actuators

To illustrate the basic motions of a two-way actuator, a simple two-way actuator is shown in Fig. 11.5 consisting of a dead weight load acting as a bias force on an SMA spring. The deflection of the spring under the bias force varies with temperature, with δ_H the deflection at high temperatures and δ_L the deflection at low temperatures. As the temperature is raised and lowered the

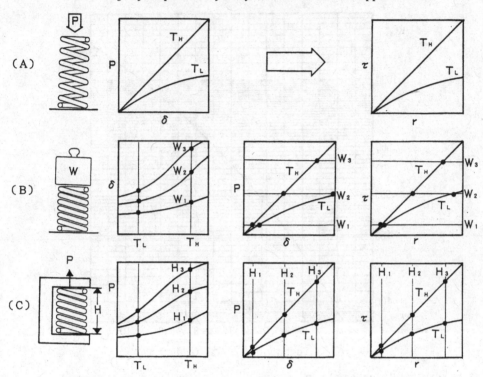

Fig. 11.4. The three methods for determining the relationship between load and deflection for an SMA spring. (A) Load vs. deflection curve at constant temperature. (B) Temperature vs deflection curve under constant load. (C) Temperature vs load under constant deflection.

interface moves between A and B and the device stroke is the difference $\delta_L - \delta_H$. Since a dead weight is not easily incorporated into a device, the bias force is usually supplied by a spring which can be fabricated from any suitable alloy.[4]

Figure 11.6(a) illustrates a typical bias-type two-way actuator in which an SMA spring and a bias spring are arranged so that they oppose each other's motion. Figure 11.6(b) shows the relationship between the load deflection curves and the two-way movement of the actuator. Since the two springs work in opposition, the slopes of the curves have opposite signs. The distance between the basal points A and B on the curves corresponds to the total compressed length as installed in the actuator. The opposing spring forces are balanced where the bias spring deflection curve crosses the SMA deflection curve, thus, at an elevated temperature the springs are balanced at point A and at low temperature at point B. The two-way motion is observed as a stroke of generated D as the temperature is raised and lowered between these two operating points. If the actuator operates against an external force the stroke is

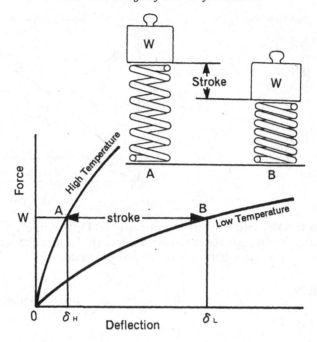

Shape memory spring behavior

Fig. 11.5. Schematic diagram of a simple two-way SMA actuator using a dead weight bias.

proportionally shortened. For example, if a load P_1 is applied at high temperature, the deflection at that temperature shifts to C and the stroke is reduced to D_1.

Since thermally actuated valves are a major application for SMA actuators, such as shown in Fig. 11.7(a), it is useful to illustrate a graphical design based on this type of mechanism. For the valve illustrated we have the following specifications:

Valve diameter: 5 mm   Fluid pressure: 0.2 MPa   Stroke length S: 3 mm

The design parameters based on Ti–Ni SMA:

τ_{max}: 150 MPa γ_{max}: 1.0% G_L: 7500 MPa G_H: 23 000 MPa   C: 6

The design procedure is as follows:

The load P_1 on the valve spring is equal to the product of fluid pressure and the cross-sectional area: $P_1 = \pi r^2 \times$ Fluid $P = 3.1416 \times 2.5^2 \times 0.2 = 3.927$ N. Figure 11.7(b) shows the shear stress *vs* shear strain for an SMA spring and for a bias spring of selected stiffness. Approximate values for the parameters G_L and G_H are obtained by measuring the slope of the SMA curve

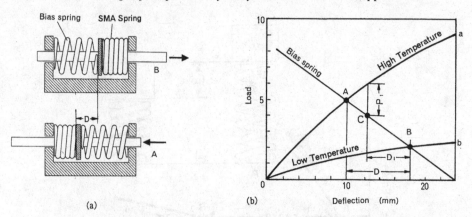

(a)

(b)

Fig. 11.6. Two-way SMA actuator with a spring bias. (a) Deflection of a two-way SMA spring with spring bias at high and low temperatures. (b) Load *vs* deflection diagram for designing a two-way SMA actuator.

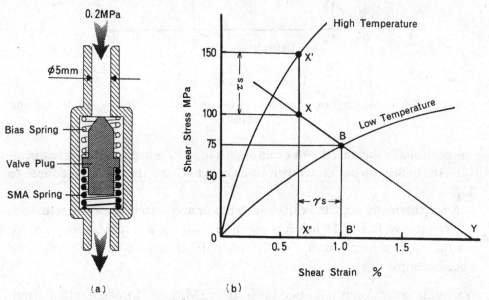

(a)   (b)

Fig. 11.7. An example of a practical design using a two-way SMA actuator. (a) Cross-section of the SMA actuated valve. (b) Stress *vs* strain diagram for the graphical design of the two-way valve actuator.

assuming that it is a straight line. The first step is to draw the stress–strain curve of the bias spring. Although this can be any chosen stiffness the bias spring is usually chosen to be roughly equal in stiffness to the SMA spring at low temperature. The bias spring line intersects the low temperature SMA stress–strain curve at a value corresponding to γ_{max} which was specified as 1%. The second step is to draw a vertical line X'X" that is used to calculate the output properties of the actuator, that is, the SMA spring in combination with the bias spring. The actual position of this line is arrived at experimentally; when the output is high strain and low stress the line is drawn near the point A and when the output is high stress and low strain the line is drawn near point B. In this example the deflection at high temperature γ_H, is defined by the line OX", 0.65%, and the stress τ_s is defined by the line X'X which is 50 MPa. The shear strain, γ_H ($= 1.0 - 0.65 = 0.35\%$) corresponds to the selected stroke length, S, of 3 mm, and the shear stress τ_s corresponds to the force P_1.

Using the previous equations:

$$\kappa = (4 \times 6 - 1)/(4 \times 6 - 4) + 0.615/6 = 1.2525$$

$$d^2 = \frac{8PC\kappa}{\pi\tau}$$

$$= 8 \times 3.927 \times 6 \times 1.2525/3.1416 \times 50 = 1.503$$

$$d = 1.226 \text{ mm which is rounded to 1.2 mm.}$$

The mean spring diameter $D = Cd = 6 \times 1.2 = 7.2$ mm
The number of active coils,

$$n = \frac{Sd}{\pi\gamma_s D^2}$$

$$= 3 \times 1.2/3.1416 \times 0.0035 \times 7.2^2 = 6.315.$$

The calculated dimensions of the SMA spring are:

Wire diameter 1.2 mm, mean spring diameter 7.2 mm with 6.3 active coils. The next step in the design is to calculate the dimensions of the bias spring. Since the values of the stress and strain shown in Fig. 11.7(b) are calculated based on the sizes of the SMA spring, they do not directly apply to the calculation of the bias spring since they are converted from the values for load and deflection. Considering the linear relationship, the maximum shear strain of the bias spring X"Y ($= 0.0145$) may be converted to the maximum deflection:

$$\delta_{max} = \frac{S \times X''Y}{B'X''} = 3 \times 0.0145/0.0035 = 12.42.$$

The maximum stress XX"($= 100$MPa) may be converted to the maximum load:

$$P_{max} = \frac{P_1 \times X'X''}{XX'} = 3.927 \times 100/50 = 7.85.$$

The spring constant for the bias spring is then:

$$K = \frac{P_{max}}{\delta_{max}} = \frac{7.85}{12.42} = 0.632 \, \text{N/mm}.$$

The stainless steel selected as the material of the bias spring has a shear modulus of 67 570 MPa and a maximum stress of 340 MPa. From the valve dimensions we observe that the bias spring mean diameter is 8 mm. The stress correction factor is assumed to be 1.2. Substituting these values into Eq. (11.3) we can calculate the value for the wire diameter:

$$d^3 = \frac{8P_{max}D\kappa}{\pi\tau} = 8 \times 7.85 \times 8 \times 1.2/3.1416 \times 340 = 0.564$$

$d = 0.82$ which is rounded to 0.8 mm.

Active coils:

$$n = \frac{Gd^4\delta_{max}}{8P_{max}D^3} = 67\,570 \times 0.8^4 \times 12.42/8 \times 7.85 \times 8^3 = 10.69 \text{ coils}$$

The specification for the bias spring is:

wire diameter d: 0.8 mm   mean diameter D: 8.0 mm   active coils n: 10.7

The operating characteristics of the actuator are directly affected by the spring constant of the bias spring; it can modify the temperature response, the available force and the hysteresis. Figure 11.8(a) shows the load *vs* deflection curves for two actuators with bias springs of different stiffness; the spring constant of bias spring A is greater than that for bias spring B. Since a strong bias spring has a large stress gradient, that is, the slope of the force–deflection curve, the length of the actuator stroke is reduced in proportion to that gradient. This decrease in stroke as a function of bias spring stiffness is illustrated in Fig. 11.8(b); for this reason, in the design of an actuator, the spring constant should be as small as possible.

Although the transformation temperature of the SMA spring is a function of the composition and the alloy processing, the operation temperature of an actuator can be modified by changing the biasing force.[5] Figure 11.9(a) illustrates a two-way SMA actuator with a variable bias. The biasing force can be adjusted by means of the adjusting screw which changes the length of the bias spring, and therefore its force. Figure 11.9(b) shows the effect of varying the biasing force on the force *vs* deflection curves and Fig. 11.9(c) shows the corresponding force *vs* temperature curves. As the compression of the bias spring increases from A to B to C, the biasing force increases proportionally and the deflection *vs* temperature curve is shifted to the higher temperature side. This means that in order to achieve a given output force the actuator must be heated to a higher temperature as the biasing force is increased. If the deflection is limited to the region to the left side of the point L, deflections

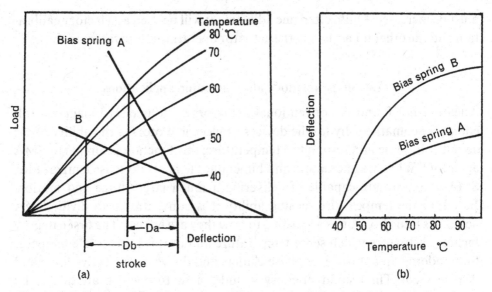

Fig. 11.8. Influence of the bias spring stiffness on the deflection characteristics of an SMA actuator. (a) Load *vs* deflection curves with two different bias springs. (b) Actuator deflection *vs* temperature curves with different bias springs.

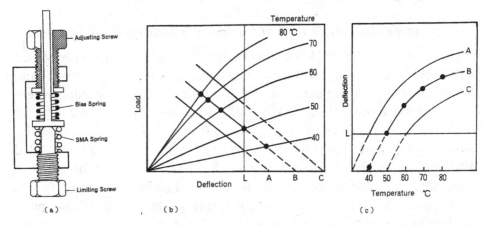

Fig. 11.9. Control of the operating temperature of a two-way actuator using variable bias force. (a) A two-way actuator with bias-force adjustment. (b) Influence of bias preload on the balancing points of the SMA spring and the bias spring – the vertical line L indicates the deflection limit. (c) Deflection *vs* temperature curve of the actuator – the horizontal line L indicates the deflection limit.

indicated by a broken line are not permitted; a condition which is typical in an actuator where there is a maximum permissible shear strain in the SMA spring. The operating temperature of the actuator can be changed by adjusting the bias screw; it can also be changed by shifting the datum point by means of the limiting screw, also shown in Fig. 11.9(a). If the screw is shortened then the temperature which is required for the actuator to develop force and deflection must be increased. Quite often one adjustment will be used as a factory calibration and the other for adjusting the actuator operating temperature in the field.

11.4 Shape memory alloy actuator applications

A feature not found in conventional actuators such as electromagnetic solenoids, pneumatic or hydraulic devices or electric motors is the ability of the actuator to sense and respond to temperature; this function is unique to SMA actuators. While wax actuators and bimetallic elements do sense temperature, SMA actuators are capable of delivering a much larger force and motion. Based on the temperature sensing ability it is conventional to divide SMA actuators into two groups according to how they are heated. The first category includes actuators which sense temperature and at the appropriate temperature undergo an appropriate phase change and deliver a corrective force and displacement. The second category includes actuators which are heated by fluid, air or electric current and caused to perform their function on demand. These two types of SMA actuators are described with examples.

11.4.1 Applications where the SMA actuator is both sensor and actuator

Control systems which employ a temperature sensing function usually consist of three elements: the temperature sensing device, an electronic processor and the electrically energized actuator. An example is the air conditioner vent controller, which will be described in detail later, consisting of a semiconductor thermistor, a microprocessor or IC circuit and an electric motor. This combination requires three components and wiring, resulting in significant added fabrication costs. By contrast, these same functions can be performed by a single SMA actuator which requires no electrical supply and has only two moving parts, the SMA spring and the bias spring. This results in lower parts and fabrication costs and an enhanced reliability. A variety of commercial thermal sensing SMA actuators are described.

Thermal protection device for a domestic water filtration unit

The use of water filtration in the home has increased and most filters are damaged if they are exposed to hot water. For this reason an SMA actuator has been designed which protects the filter from accidental exposure to hot water where the faucet is a mixing type which has water temperature ranging from cold to hot. Figure 11.10 illustrates the operating principle of the thermal protection valve for filter protection.[6] When hot water is flowing the SMA spring is heated, overcoming the bias spring force, and closes the valve so that

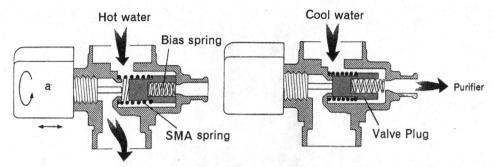

Fig. 11.10. Operating principle of thermal protection valve. (Left) Bypassing hot water. (Right) Passing cold water to the filter.

the hot water cannot flow through the filter but is directed to the faucet. When cold water flows the bias spring pushes the valve away from the seat and the cold water flows through the filter. If bypassing the filter is required the bias spring can be disabled by turning the handle (a) so that the cold water flows directly through the faucet.

Steam trap for passenger train steam heating system

Trains usually use circulating steam for heating the passenger or freight cars which requires the steam to flow from the steam generator, usually located in the diesel locomotive, through a piping system through heat exchangers in the cars and back to the generator. Steam condenses in the circuit, and unless this condensate is drained away by a steam trap the steam flow is blocked. Gas sealed metal bellows operated steam traps have been widely used for this purpose but suffer from the fact that when operated in the winter the conden-sate can freeze and cause the bellows to rupture. The SMA operated steam trap shown in Fig. 11.11 avoids this problem.[7] As condensate water accumulates in the trap, the temperature of the trap gradually falls and when the temperature of the SMA spring reaches the M_f the bias spring causes the trap to open and drain the accumulated condensate. With the valve open, steam can immediate-ly flow through and heat the SMA above the A_f causing the valve to close. The opening and closing response times of the valve operating on the actual train system with an ambient temperature of $-10°C$ are shown in Fig. 11.12. The valve filled with condensate in 3.6 minutes and the closing time when the SMA spring was heated by the steam flow was 4.2 seconds.

Thermally actuated electric switch for diesel engine radiator fan

The cooling system in a diesel engine employs an electrically driven fan which is turned off and on as required to maintain a specified system temperature.

Fig. 11.11. Steam trap for the heating system in a diesel train.

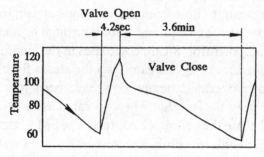

Fig. 11.12. Temperature *vs* time graph of the steam trap valve in operation.

The thermally operated electrical switch must operate reliably while subjected to strong vibrations. Conventional bimetal operated switches used in this application suffer from chatter due to their low contact force and short stroke. These conditions have been completely corrected by the use of the high contact force SMA actuated switch, shown in Fig. 11.13.[7] Figure 11.14 shows the basic structure of this type of thermal switch. Although the trend to miniaturization of electrical components is strong, there is a large demand for switch actuators which have large force and large stroke capability; SMA actuators are particularly well suited for this type of device.

Fig. 11.13. SMA actuated thermal switch in a diesel engine cooling system.

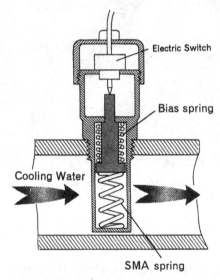

Fig. 11.14. Cross-section of the diesel engine cooling system thermal switch.

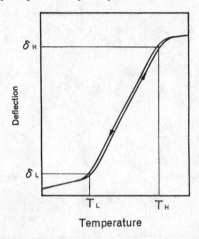

Fig. 11.15. Deflection *vs* temperature curves for the linear shape memory component used in the mixing valve.

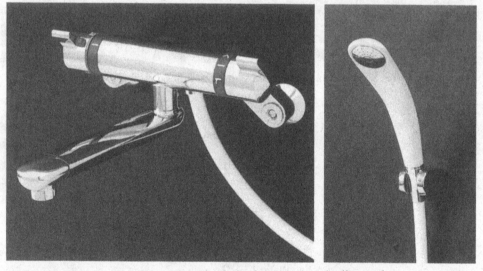

Fig. 11.16. Appearance of the thermal mixing faucet using the linear shape memory actuator shown in Fig. 11.17. (Courtesy of TOTO Ltd[8])

Thermostatic mixing valve

In the SMA thermal actuators described above the action is basically an on-off motion at a particular temperature or temperature range. When used as a thermostatic device, linear motion with minimum hysteresis is required, and a deflection *vs* temperature curve for such a valve in operation is shown in Fig. 11.15. This SMA device operates in a linear manner over the temperature range of T_L to T_H.

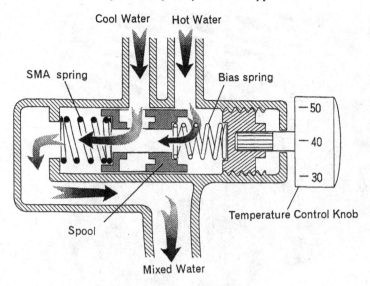

Cool Water   Hot Water

SMA spring   Bias spring

— 50
— 40
— 30

Temperature Control Knob

Spool

Mixed Water

Fig. 11.17. Schematic of the thermal mixing valve using the linear shape memory spring.

Figure 11.16 shows a thermostatically controlled faucet which mixes hot and cold water and delivers water at a controlled temperature.[8] The operating principal of the thermostatic mixing valve is illustrated in Fig. 11.17. An SMA spring, opposed by an adjustable bias spring, is exposed to the flow of mixed hot and cold water. When the water temperature is too hot the SMA spring pushes the spool to the right, restricting the flow of hot water and lowering the mixture temperature. The reverse motion occurs when the water is too cold. The delivery temperature of the water is adjusted by the control knob which changes the compression on the bias springs.

Air flow control for an air conditioner

The temperature of the air flow from an air conditioner is controlled by switching the compressor or changing the power frequency. The exhaust air from the air conditioner is constantly changing and can sometimes be uncomfortably cool. To avoid this problem, control of the air flow direction is effected by an SMA actuated moving flap mounted in the front of the air conditioner shown in Fig. 11.18.[9] When the air flow temperature is cold the flow is directed in an upward direction, and when the air temperature exceeds body temperature the air flow is directed downwards. The link mechanism and the rocking motion of the flap are illustrated in Fig. 11.19 which shows the SMA spring and bias spring operating through a pivot to swing the flap seesaw-like to direct the air flow direction.

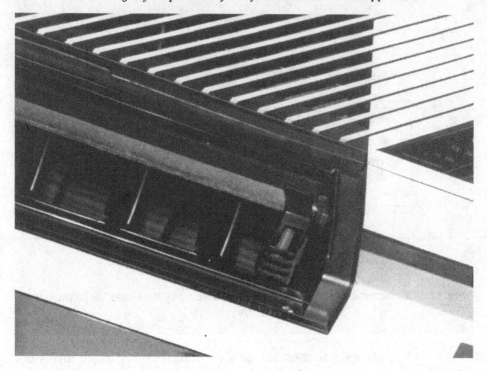

Fig. 11.18. Moving flap for directing air flow in a room air conditioner.

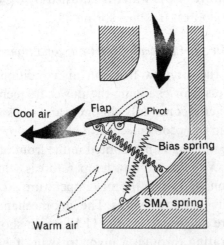

Fig. 11.19. Schematic of the link system flap control for air flow control.

11.4.2 Applications as electrically heated SMA actuators

Although there is an advantage to the use of actuators which combine the functions of temperature sensing with actuation, there are many cases where they are used simply as actuating units. In these cases the SMA element is heated by some external means such as an electric current, hot fluid or hot air or even radiant heat from a light source.

Air damper for a multi-function electric oven

Multi-function ovens which can operate as either a microwave oven or a circulating hot-air convection oven depending on the food being prepared and the desired cooking method are now available. These ovens require different modes of air flow control; they must be sealed during convection operation but must be ventilated when operating in the microwave mode in order to expel vapor from the oven cabinet. A damper which can open and close to regulate the air flow has typically been operated by an electric motor. A new SMA damper control offering simplicity and space savings is shown in Fig. 11.20, where it is compared with the older style magnetic drive.[10]

SMA operated fog lamp protective louver

Fog lamps are mounted in the front of a car close to the road and in this position are vulnerable to damage from stones. The damage can be minimized if when not in use the lamps are protected by a louver shutter which is opened and closed by an SMA actuator. Figure 11.21 shows a fog lamp with shutters

Fig. 11.20. SMA bias spring combination for multi-function electric oven vent control (right) compared with a conventional electric control (left).

Fig. 11.21. Car fog lamp with SMA actuated protective shutters.

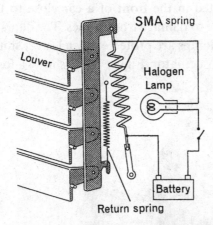

Fig. 11.22. The operating mechanism of the car fog lamp protective shutters.

that are opened and closed by an SMA actuator when the lamp is turned on and off.[11] The mechanical arrangement and electrical circuit are illustrated in Fig. 11.22. The shutter system is electrically connected in series with the lamp so that when the lamp is turned on the SMA spring is heated and the shutter opens. When the lamp is turned off the bias spring closes the shutter and holds it in this position.

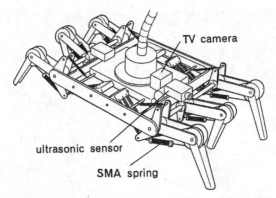

Fig. 11.23. The six leg robotic crab. (After Furuya *et al.*[12])

Robotic crab with six legs

Underwater exploration and geological surveying has become important and there is a constant search for unmanned robotic devices to carry out these missions. A six leg SMA actuated robotic crawler by Furuya *et al.* has the potential for meeting the requirements of this type of operation. A schematic of the prototype robot is shown in Fig. 11.23.[12] The Ti–Ni SMA has excellent sea water corrosion resistance and since the electrical resistance of Ti–Ni is much lower than sea water the actuators can be electrically energized without current leakage into the surrounding water. The fact that the actuator wires can be bare means that they are cooled very effectively by the water and, as such, their cycle times are quite short, even for high power actuators using heavy SMA wires. Conventional undersea devices using electrical or hydraulic actuators must be sealed to prevent sea water leakage, a difficult task when operating in deep water; the bare SMA actuators need no such protection.

SMA ribbon actuators for robotic hands

As we have pointed out earlier, the limiting speed of an SMA actuator is the cooling part of the cycle. For robotic devices such as the hand illustrated in Fig. 11.24,[13] the actuator elements must be cooled by air which limits the speed of the hand motion. By using flat ribbon SMA actuator elements with their larger surface area the cooling rate is enhanced resulting in an increase in the speed of the robotic finger motion by 20%.

11.4.3 The applications of SMA in micro actuators

Micro machines are not simply miniature versions of conventional machines but are novel mechanisms based on new design concepts and new fabrication

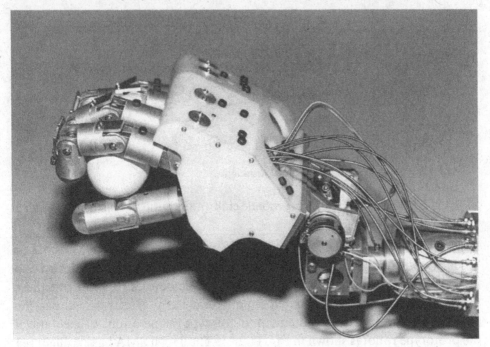

Fig. 11.24. Robot hand driven with multiple SMA flat wires.[13]

techniques. They are the natural interface with the growing world of micro electronics and can be thought of as intelligent miniature machines. Some of the future applications could include devices which can be inserted in the human body to carry out diagnostic or repair functions. Since a micro machine requires a micro actuator capable of generating force and motion with a minimum energy input, actuator development has centered on SMA devices using micron sized wires and thin films. The simplicity of SMA actuators with their unique flexibility and variety of motion make them good candidates for micro machine actuators. In addition, the cooling limitation which is imposed on more conventional SMA actuators is less of a problem with micro actuators because of their large surface to weight ratios. Several prototypes of micro actuators are described below.

Mollusk-type manipulator for transplanting saplings

Transplanting saplings in biotechnology studies demands great skill on the part of the operator. Conventional robotic devices have not proven useful in this application because of an inability to handle the weak sapling without damage. Figure 11.25 shows a robotic manipulator developed by Miwa[14] which uses three fingers in the front end of a mollusk-like structure. The SMA

Fig. 11.25. Mollusk-type manipulator for sapling transplanting. (After Miwa[14])

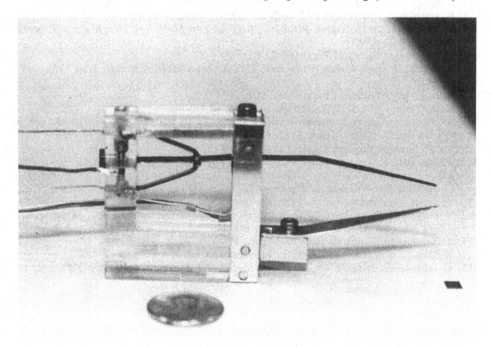

Fig. 11.26. Photograph of the differential SMA actuator driven miniature clean
gripper. (After Ikuta[15])

wire fingers act as both structural members as well as the moving manipulators and can pick up a sapling and after inspection transplant it to a new culture.

Miniature clean gripper

A miniature gripper for use in clean rooms or in vacuum has been demonstrated by Ikuta[15] and is shown in Fig. 11.26. The device consists of a pair of cantilever phosphor bronze fingers which have SMA springs mounted above and below their surfaces so that when one side is energized the fingers move together to effect a gripping action. The SMA actuators in this type of geometry are called 'differential actuators' and since the moduli of the upper and lower springs are out of phase, this type of actuator can produce a force and stroke superior to a conventional biased actuator. The motion of the grippers is detected by a photo diode and light emitting diode couple which makes possible accurate detection of the gripper position. Since there are no sliding parts or seals this device is particularly suited to operations where contamination is a problem.

References

1. H. Carlson, *Spring Designer's Handbook*, (Marcel Dekker, New York, 1978).
2. Y. Suzuki and H. Horikawa, *MRS Int'l. Mtg. on Adv. Mats*, Vol. 246, (Material Research Society, 1992), p. 389.
3. Kato Hatsujo Technical Report 90-D-002.
4. Y. Suzuki, in *Shape Memory Alloys*, ed. H. Funakubo (Gordon Breach, New York, 1987), p. 181.
5. Kato Hatsujo Technical Report 92-S-001.
6. Kato Hatsujo PRODUCT GUIDE 1993-289.
7. I. Ohkata, Advanced Materials '93, V/B: Shape Memory Materials and Hydrides (ed. K. Otsuka and Y. Fukai), *Trans. MRS Jpn.* **18B** (Elsevier Sci., B.V., 1994), p. 1125.
8. Toto Plumbing Fixtures Export Catalog '94–95.
9. T. Todoroki, K. Fukuda and T. Hayakumo, *Kogyo-Zairyo* (in Japanese), **32**, no. 7, (1984), 85.
10. H. Tanaka, *Denshi-Gijutu* (in Japanese), **26**, no. 2, (1983), 46.
11. Y. Suzuki, *System and Controls* (in Japanese), **29**, no. 5, (1985), 296.
12. Y. Furuya, H. Shimada, Y. Goto and R. Honda, in *Engineering Aspects of Shape Memory Alloys*, ed. T. W. Duerig *et al.*, (Butterworth–Heinemann, London, 1990), p. 599.
13. Y. Yamamoto, *Furukawa Review*, (March, 1986), p. 39.
14. Y. Miwa, *Seimitsu Kougakkaishi* (in Japanese), **53**, no. 2, (1987), 15.
15. K. Ikuta, *Proc. of IEEE Int. Conf. on Robotics and Automation*, (1990), p. 2156.

Medical and dental applications of shape memory alloys

S. MIYAZAKI

12.1 Introduction

Among many materials including metals, alloys, ceramics etc. available commercially, only a limited number are currently being used as prostheses or biomaterials in medicine and dentistry. The reason for this is that prostheses need to satisfy two important demands, i.e., (a) biofunctionability and (b) biocompatibility.[1,2] Biofunctionability refers to the ability of the biomaterials to perform the desired function for the expected period in the body. Biocompatibility is the ability of the materials to be nontoxic during the implanted period. Because of these rigorous demands, only the following three metallic materials have been qualified to be available as implant materials, i.e., Fe–Cr–Ni, Co–Cr and Ti–Al–V. However, shape memory alloys have been recently introduced to medicine and dentistry, since they have unique functions such as shape memory effect, superelasticity and damping capacity. Among several tens of shape memory alloys, Ti–Ni alloy is considered to be the best because of its excellence in mechanical stability, corrosion resistance, biofunctionability, biocompatibility, etc. The purpose of this chapter is to introduce some clinical application examples using the Ti–Ni shape memory alloy and to introduce some data on corrosion resistance, elution testing and biocompatibility of the alloy.

12.2 Application examples

Dental applications

Unique functions such as shape memory effect and superelasticity possessed by Ti–Ni alloys[3] have recently been used successfully in dental and medical fields as well as in engineering fields. In dentistry, a highly successful application example is a dental arch wire[4–9] which uses the superelasticity characterized by

267

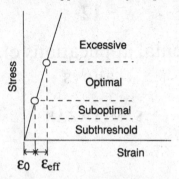

Fig. 12.1. Schematic stress–strain curve of an elastic orthodontic wire with a high elastic modulus. Effective strain range corresponding to the optimal force zone is small. (After Miyazaki *et al.*[8])

a constant stress upon loading and unloading. Other interesting application examples are dental implants[10] and attachments for partial dentures[11] which use the shape memory effect. These three examples are selected to show the advantages of using Ti–Ni shape memory alloys in dentistry in the following.

Historically, a number of metals and alloys have been used for orthodontic arch wires. If a high elastic modulus alloy such as stainless steel is used, an effective strain range (ε_{eff}) corresponding to an optimal force zone is very small as shown in Fig. 12.1. In the optimal force zone, the nature of the applied force is such that it will encourage the optimum biological response, i.e., the bone remodeling response is most efficient. Above the optimal force zone is located the excessive force zone where tissue damage may occur. Below the optimal zone lie suboptimal and subthreshold force zones. In this region teeth move less efficiently and may even come to a complete standstill if forces become minimal. If a low elastic modulus alloy is used, the effective strain range expands, hence clinically the optimal force zone becomes wider. Therefore, a low elastic modulus alloy provides a greater range of activation, and fewer adjustments need to be made to the arch wire to move the teeth to their final positions. Work-hardened Ti–Ni wires have been successfully used as such low elastic modulus materials.[12,13] The goal of the above approach using lower modulus materials is to obtain a zero modulus material or a constant-force material which introduces a new orthodontic treatment.

Recently, superelastic Ti–Ni alloy was introduced to orthodontics[4–9] based on the discovery of superelasticity in Ti–Ni alloy.[14–16] This alloy exhibits unique and ideal mechanical properties for the practice of orthodontic mechanotherapy. The superelasticity in Ti–Ni alloys was first reported in 1981.[15,16] The basic concept for developing stable superelasticity was presented in 1982,[14] making it possible to control the superelasticity characteristics in Ti–Ni alloy. Figure 12.2 shows the stress–strain curve of the superelastic

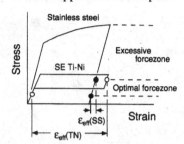

Fig. 12.2. Schematic stress–strain curves representing the characteristics of stainless steel and superelastic Ti–Ni wires. Effective strain range corresponding to optimal force zone is larger in the Ti–Ni wire than in the stainless steel wire. (After Miyazaki et al.[8])

Ti–Ni wire compared with that of a stainless steel wire. It can be seen that the effective strain range corresponding to the optimal force zone is ideally wide in the superelastic wire. The most important characteristic in the superelastic behavior is the generation of a constant force and a large reversible deformation over a long activation span. Besides, the constant force can be adjusted over a wide range by changing manufacturing and metallurgical factors such as thermomechanical treatment, composition, etc.[17] It is apparent that superelasticity presents the orthodontists with better mechanical characteristics when compared with conventional elastic materials such as stainless steel. Figures 12.3(a) and (b) show a clinical example of orthodontic treatment using a superelastic Ti–Ni arch wire. The result showed faster movement of teeth and shorter chair time when compared with the case of a stainless steel wire.

The second example is teeth-root prosthesis. Among several methods which restore the mastication function of patients missing more than one tooth, teeth-root prosthesis is considered to be the method which creates the most natural mastication function. Blade-type implants made of Ti–Ni shape memory alloy have been used in Japan, the examples being shown in Fig. 12.4. Blade-type shape is considered to be suitable for applying to the Japanese who possess comparatively narrow jaw-bone structure.[10] Since the open angle of the blades is changeable by heating the shape memory alloy implant, the implant is set to ensure tight initial fixation and to avoid accidental sinking on mastication with an open blade shape, keeping the condition of insertion of the implant to the jaw-bone easy with a closed blade shape. The reverse martensitic transformation finish temperature A_f is set to be 40°C so that the implant will open the blades after setting and warming up to 42°C by warm physiological saline.

Another application of the shape memory alloy in dentistry is for partial dentures. The key to the partial denture is the development of an attachment

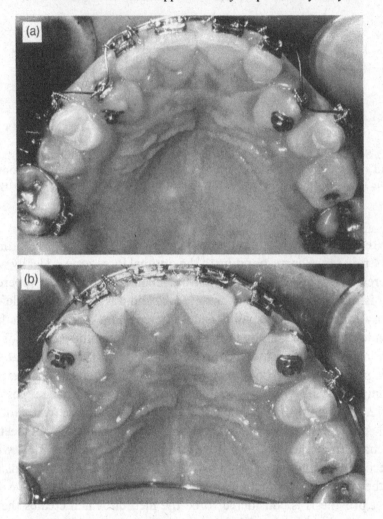

Fig. 12.3. Orthodontic treatment using a Ti–Ni superelastic arch wire. (a) Malaligned teeth before treatment and (b) normally aligned teeth after the first stage of treatment. (After Sachdeva and Miyazaki[7])

for connecting the partial denture with retained teeth.[11] Clasps have been conventionally used for about a century as the attachment for a partial denture. One of the defects of clasps made of conventional elastic materials is a loosening during use; this can be improved by replacing the elastic materials by a superelastic Ti–Ni alloy. However, the clasps have an esthetic problem, because they are visible along the teeth alignment. In order to solve the problem, the size of the attachment must be smaller than the width of the teeth; such a precision attachment using a small screw has recently become available. However, they have to be designed and fabricated very precisely so that they

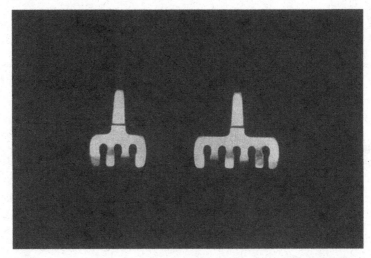

Fig. 12.4. Blade-type implants made of Ti–Ni shape memory alloy. (After Fukuyo *et al.*[10])

lack flexibility in order to absorb an external shock and follow the change in setting condition due to the shape change of a jaw-bone during long term use. Since the shape memory alloy itself has a function as a joint because of its flexible nature, if it is used for the attachment of a partial denture, there is the possibility of solving the above problems existing in clasps and precision attachments. It also becomes possible with such shape memory attachments to easily attach and detach the partial dentures.

The attachment consists of two parts, i.e., one is a fixed part which is made of a conventional dental porcelain–fusible cast alloy and attached to the full cast crown on anchor teeth, while the other is a movable part which is made of the Ti–Ni shape memory alloy and fixed on the side of a partial denture.[11] Examples of the movable part of the attachment are shown in Fig. 12.5. Figure 12.6 shows the fixed part attached to the cast crown on anchor teeth. The cast crowns were surface treated to be shiny and coated with porcelain teeth caps.

Medical applications

Many clinical applications of Ti–Ni shape memory alloy have also been reported in medicine. Among many medical fields, orthopedics is one of the most attractive fields for applications of Ti–Ni shape memory alloy. Bone plates are attached with screws for fixing broken bones. Bone plates made of Ti–Ni shape memory alloy will be more effectively applied for connecting the broken bones than conventionally used material bone plates,[18–20] because the shape memory alloy bone plates can provide compressive force on the fracture

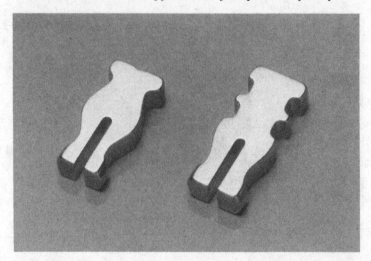

Fig. 12.5. Movable part of the shape memory alloy attachment for partial denture. (After Miyazaki *et al.*[11])

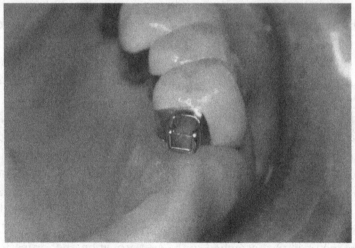

Fig. 12.6. Fixed part of the shape memory alloy attachment for partial denture. (After Miyazaki *et al.*[11])

surfaces of the broken bones as well as a fixation function as shown in Fig. 12.7: healing proceeds faster under uniform compressive force. U-shaped staples made of Ti–Ni shape memory alloy are also available for joining broken bones.[21]

A marrow needle is used for the fixation of a broken thigh bone. The shape and size of the needle need to be complicated and larger than the interior of the marrow in order to reinforce the broken bone. This demand makes it difficult to insert the needle into the broken bone. Using the shape memory effect, insertion will be greatly improved as shown in Fig. 12.8, without loosing the

SMA bone plate   Bone

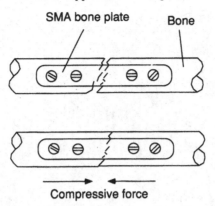

Compressive force

Fig. 12.7. Schematic diagram showing the advantage of a shape memory alloy bone plate. (After Onishi *et al.*[20])

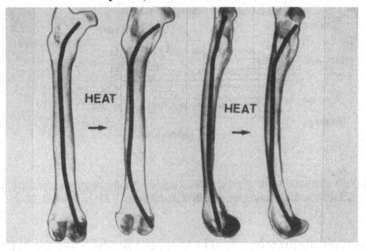

Fig. 12.8. Marrow needles showing their shapes before and after heating. (After Onishi *et al.*[20])

reinforcing function, because the necessary size and shape will recover after heating the needle in the marrow.

Currently available joint prostheses are made of bone cement to be fixed in the bone. Stress acting on the joint prosthesis is quite intense and severe; 3–6 times the body weight of the patient under nominal action and such stress being cycled up to 10^6 times. Conventional bone cement causes several inconveniences: gradual loosening after implantation and resultant infection and other complications. To avoid such problems, a prosthetic joint made of Ti–Ni shape memory alloy was developed.[20] High wear resistance is also another advantage of using a Ti–Ni prosthetic joint.

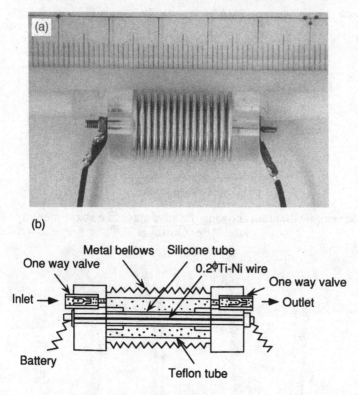

Fig. 12.9. (a) An artificial kidney pump utilizing Ti–Ni shape memory alloy wires and (b) the corresponding schematic cutaway diagram. (After Sekiguchi *et al.*[22])

Another application is the spinal bent-calibration bar (so-called Harlington bar). One of the problems with a conventional bar is the loss of elastic force after an operation, hence re-operations are necessary for calibrating the force of the bar. A Ti–Ni bar will not loose the elastic force so easily and the number of re-operations will be decreased.

Shape memory alloys can also be used for artificial organs such as actuators for driving an artificial kidney pump[22] and an artificial heart pump.[23] For these applications, high fatigue strength and miniaturization are demanded. Ti–Ni alloy can fulfill both demands. Figures 12.9(a) and (b) show a prototype of an artificial kidney pump and the schematic cutaway diagram illustrating the mechanism of the pump. The function of the artificial kidney pump is to take out blood from an artery and send it back to a vein after subjecting the blood to a variety of treatments. In the treatments it is necessary to supply a small amount of anticoagulant at a rate of only about 50 μl per minute. Moreover it is necessary to develop a micropump in order to miniaturize the whole system of the artificial kidney. There will be the possibility of solving these problems using a shape memory alloy.

As for miniaturization technology, a new application field for shape memory alloys has been growing since 1987. The purpose of the new field is to fabricate micromachines or microrobots which are considered to be a revolution in future technologies including engineering and medicine. In order to fabricate such micromachines and microrobots, many essential technologies need to be developed. One of the necessary technologies is that for developing microactuators which can power such micromachines. Among several candidate materials for such microactuators, shape memory alloy thin films have great advantages such as large force and deformation. Besides, a great disadvantage of shape memory alloys, i.e., a slow response due to the limitation of cooling rate, can be greatly improved by miniaturization. Very recently, a perfect shape memory effect has been developed in sputter-deposited Ti–Ni thin films of the order of μm in thickness.[24,25] In the future, there will be possibilities for fabricating micromachines which work in our bodies for microtreatments and microsurgery with the assistance of microactuators made of Ti–Ni shape memory alloy thin films.

One of the most sophisticated medical applications of shape memory alloys is an active endoscope.[26] Endoscopes are indispensable and commonly used in medical and industrial fields. One of the problems with endoscopes is the difficulty of insertion when the path is narrow and complicated. In order to abate the discomfort of patients during insertion, the endoscopes should be flexible and controllable. Significant improvement in such characteristics was made in an endoscope using shape memory alloy actuators which enable each joint of the endoscope to possess the function of all-round bending. Figure 12.10 shows the prototype of such an endoscope which has several joints driven by shape memory alloy actuators. A microcomputer controls the bending angle of each joint by watching the shape of the path during insertion. Since the microcomputer can store the memory of the shape of the path, the second and other following joints can be controlled in order to accommodate the whole shape of the endoscope to the shape of the path.

Superelasticity has also been used successfully in medical fields.[27] One of the typical applications is a guidewire which is used as a guide for safe introduction of various therapeutic and diagnostic devices. The guidewire is a long, thin metallic wire which is inserted into the body through a natural opening or small incisions. Stainless steel has long been used almost exclusively for guidewires. However, superelastic Ti–Ni alloy has recently been replacing stainless steel for guidewires. The anticipated advantages of using superelastic Ti–Ni alloy instead of using stainless steel include (a) diminishing the compli-

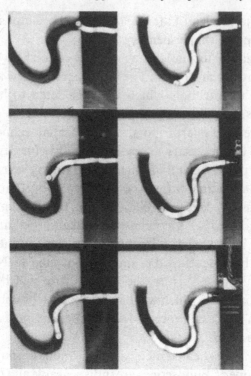

Fig. 12.10. Sequential motion of the active endoscope made of shape memory joints controlled by a microcomputer. (After Hirose *et al.*[26])

cation of the guidewire taking a permanent kink which may be difficult to remove from the patient without injury, and (b) increased steerability to pass the wire to the desired location due to an ability to translate a twist at one end of the guidewire into a turn of nearly identical degrees at the other end.

Because of the limited space allotted to this chapter, only some very typical examples of applications have been shown in order to explain the advantages and concepts of using Ti–Ni shape memory alloys in medicine and dentistry. Since there have been many papers published on medical applications, readers who want to know more medical aspects of the Ti–Ni shape memory alloys can refer to reference papers listed at the end of this chapter, e.g., Refs. [28–33] for application examples and Ref. [33] for corrosion and biocompatibility.

12.3 Corrosion resistance

Corrosion resistance of Ti–Ni shape memory alloy has been studied by the anodic polarization measurement method by several groups. It is necessary and valuable to compare the corrosion characteristics of the Ti–Ni alloy and

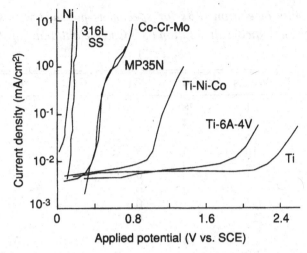

Fig. 12.11. Anodic polarization curves for Ni, 316L stainless steel, MP35N, Co–Cr–Mo, Ti–Ni–Co, Ti–6Al–4V and Ti. Potentiostatic measurements were made in Hank's physiological salt solution at 37°C and pH 7.4. (After Speck and Fraker[34])

other biomaterials. The results are summarized in the following on the basis of their description.

Figure 12.11 shows the results for Ni, 316L stainless steel, MP35N, Co–Cr–Mo, $Ti_{49.5}$–$Ni_{49.0}$–$Co_{1.5}$ (at%), Ti–6Al–4V, and Ti measured in Hank's physiological solution at 37°C and a pH of 7.4.[34] These materials, except Ni and Ti, are used in dentistry and medicine. The anodic polarization curves for these materials show the breakdown potential at which there is a rapid increase in current density. The pure Ni and stainless steel have the lowest breakdown potentials, the breakdown occurring at a voltage less than + 0.2 volts. In Co–Cr–Mo and MP35N, the polarization curves are very similar and show the breakdown at a voltage of + 0.42 volts. These materials exhibit passivity in the plateau region of the curves, because they are coated by an oxide film. The oxide film dissolves at the breakdown potential, resulting in the formation of pitting. The titanium alloys including Ti–Ni–Co (shape memory alloy) and Ti–6Al–4V have a wider passive region than the above materials and higher breakdown potentials, indicating that the titanium alloys including shape memory alloys are more corrosion resistant. The breakdown potential for Ti–Ni—Co is + 1.14 volts, while that for Ti–6Al–4V is + 2.0 volts. Pure Ti has the highest breakdown potential, which is + 2.3 volts.

In order to improve the corrosion resistance of the Ti–Ni alloy, surface coating is considered to be effective. The effect of surface coating with TiN, TiCN and oxide films has been investigated.[35] All these surface coating materials are effective in decreasing the amount of current density; it became

Table 12.1. *Results of elution tests for specimens polished with emery paper in 0.9% NaCl physiological salt solution (37°C) (weight unit: μg). (After Fujita et al.[36])*

Sample	Weeks	Ni	Ti	Cr	Fe	Mo
Ni	1	28				
	2	4				
	5	5				
	10	8				
Ti	1		0.6			
	2		0.5			
	5		0.3			
	10		0.1			
SUS	1	24		0.8	270	
(304)	2	3		0.0	6.5	
	5	3		0.0	3.2	
	10	1		0.0	10	
SUS	1	55		0.8	630	23
(316)	2	10		0.0	8.0	3
	5	6		0.0	8.8	1
	10	3		0.0	13.0	1
Ti–Ni (M)	1	10	0.8			
(M_s: 40°C)	2	2	1.1			
	5	5	1.3			
	10	3	1.3			
Ti–Ni (P)	1	2	0.3			
(M_s: 0°C)	2	1	0.2			
	5	1	0.1			
	10	1	0.1			

1/100 to 1/1000 times that of uncoated Ti–Ni alloy when measured at 2.0 volts.

12.4 Elution test

Table 12.1 shows the results of elution tests for various types of specimens polished with emery paper; they are Ni, Ti, SUS 304, SUS 316, the parent phase of Ti–Ni and the martensite phase of Ti–Ni.[36] These specimens were kept in 0.9% NaCl physiological salt solution at 37°C during the elution test. The amount of dissolution of each specimen was determined by measuring the solution. After analyzing the solution, the specimen was again immersed in new solution. The time (week) described in the table indicates the accumulation of time. For example, 2 weeks means that the corresponding elution test

Table 12.2. *Elution of Ni from (a) specimens polished with emery paper and (b) passivation-treated specimens in 0.9% NaCl physiological salt solution (at 37°C for 1 week) (weight unit: μg). (After Fujita et al.[36])*

Specimen	(a)	(b)
SUS 304	24	0.0
SUS 316	55	0.0
Ti–Ni ($M_s = 10°C$)	2	0.0

was performed during the second single week, while 5 weeks means that the corresponding elution test was performed for 3 weeks from the third week to the fifth week. A considerable amount of Ni dissolves from Ni, SUS 304 and SUS 316, while a smaller amount of Ni dissolves from Ti–Ni alloys. Interestingly, less Ni dissolves from the parent phase of Ti–Ni than from the martensitic phase. This is important and useful information, because Ti–Ni will be in the parent phase after implantation using the shape memory effect or superelasticity, indicating that the amount of Ni dissolved after implantation will be minimal. Other data also show that Ti–Ni and Ti–Ni–Cu alloys release less Ni than dental Ni–Cr alloys.[37]

Table 12.2 shows the effect of passivation-treatment on the dissolution of Ni during the first week by comparing the results of the elution tests for emery-paper-polished and passivation-treated specimens, i.e., SUS 304, SUS 316 and the parent phase of Ti–Ni.[36] The dissolution of Ni from any of these specimens was very effectively suppressed by the passivation-treatment.

12.5 Biocompatibility

Two kinds of plates made of Ti–Ni alloy and stainless steel were implanted on the surface of the femurs of mature New Zealand White rabbits for biocompatibility tests.[38] Three months after the implantation, new bone was formed markedly not only at the bone–plate interface, but also on the surface of the Ti–Ni plate. After nine months, the Ti–Ni plated bone had much callus and showed new medullary canal formation. No connective tissue was observed between the Ti–Ni plate and the bone, whereas a connective tissue layer was observed between the stainless steel and the bone. Six and nine months after the implantation, a small amount of Ni element was detected in both cases.

Fracture healing experiments were conducted under anesthetic conditions in eight Beagle dogs.[38] Their femoral shafts were osteotomized transversely followed by the insertion and reduction of Ti–Ni and stainless steel intramedullary rods. One month after the insertion, much callus was formed at the bilateral edges in the Ti–Ni cases. The Ti–Ni cases showed significantly larger callus formation than the stainless steel cases after three months. One and a half years after the insertion, inflammatory signs, degenerative change or atypical cell infiltration was observed around the Ti–Ni rods. The amount of Ni element detected by energy dispersion X-ray scan was about the same as the background signal.

When conventional metallic blade-type dental implants are used, connective tissues called periimplant membrane will be generated in many cases. However, periimplant membrane was not observed when a blade-type dental implant made of Ti–Ni shape memory alloy was applied.[39] Even after nine months from the implantation, no connective tissue was observed around the implant. Many successful clinical cases using the Ti–Ni shape memory alloy dental implants have been reported.[10] Successful applications of Ti–Ni implants have also been reported in orthopedics, especially in China[30,32] and Germany.[19,21,40] However, it is also true that data showing the possibility of biocompatibility problems have also been reported. We need to clarify the biocompatibility problems by systematic research in order to present more reliable results. On the other hand, coating technology might need to be developed in order to increase the safety of shape memory alloys.

References

1. L. S. Castleman and S. M. Motzkin, *Biocompatibility of Clinical Implant Materials*, Vol. 1, (CRC Press, 1981) p. 129.
2. J. Autian, *Artif. Organs*, **1** (1977) 53.
3. S. Miyazaki and K. Otsuka, *ISIJ International*, **29** (1990) 353.
4. Y. Ohura, *J. Jpn. Orthod. Soc.*, **43** (1984) 71.
5. C. J. Burstone, B. Qin and J. Y. Morton, *Am. J. Orthod.*, **187** (1985) 445.
6. F. Miura, M. Mogi, Y. Ohura and H. Hamanaka, *Am. J. Orthod. Dentofac. Orthop.*, **190** (1986) 1.
7. R. Sachdeva and S. Miyazaki, *Proc. MRS Int'l Mtg on Adv. Mats.*, **9** (1989) 605.
8. S. Miyazaki, Y. Oshida and R. Sachdeva, *Proc. International Conference on Medical Application of Shape Memory Alloys*, Shanghai (1990) p. 170.
9. R. Sachdeva and S. Miyazaki, in *Engineering Aspects of Shape Memory Alloys*, ed. T. W. Duerig *et al.*, (Butterworth–Heinemann Ltd., 1990) p. 452.
10. S. Fukuyo, Y. Suzuki and E. Sairenji, ibid., (1990) p. 470.
11. S. Miyazaki, S. Fukutsuji and M. Taira, *Proc. ICOMAT-92*, Monterey, California (1993) p. 1235.
12. G. F. Andreasen and T. B. Hilleman, *J. Am. Dent. Assoc.*, **182** (1971) 1373.
13. G. F. Andreasen and R. E. Morrow, *Am. J. Orthod.*, **73** (1978) 142.
14. S. Miyazaki, Y. Ohmi, K. Otsuka and Y. Suzuki, *J. de Phys.*, **43**, Suppl. 12, (1982), C4-255.
15. S. Miyazaki, K. Otsuka and Y. Suzuki, *Scr. Metall.*, **15** (1981) 287.

16. S. Miyazaki, T. Imai, K. Otsuka and Y. Suzuki, *Scr. Metall.*, **15** (1981) 853.
17. S. Miyazaki, in Ref. [9], (1990) p. 394.
18. G. Bensmann, F. Baumgart and J. Haasters, *Tech. Mitt. Krupp. Forsch.*, **40** (1982) 123.
19. J. Haasters, G. v. Salis-Solio and G. Bensmann, in Ref. [9], (1990) p. 426.
20. H. Ohnishi and Jinkou Zouki, *Artif. Organs*, **12** (1983) 862.
21. F. Baumgart and G. Bensmann, *Blech. Rohre. Profile*, **26** (1979) 667.
22. Y. Sekiguchi, T. Dohi and H. Funakubo, *Proc. 8th Ann. Meet. Soc. Biomater.*, **17** (1971) 470.
23. P. N. Sawyer, M. Page, B. Rudewald, H. Lagergren, L. Baselius, C. McCool, W. Halpein and S. Shinivasan, *Trans. Amer. Soc. Artif. Int. Organs*, **17** (1971) 470.
24. S. Miyazaki and A. Ishida, *Mater. Trans., JIM*, **35** (1994) 14.
25. S. Miyazaki and K. Nomura, *Proc. IEEE Micro Electro Mechanical Systems (MEMS-94)*, Oiso, Japan, (1994) p. 176.
26. S. Hirose, K. Ikuta, M. Tsukamoto, K. Sato and Y. Umetani, *Proc. ICOMAT-86*, Nara, Japan, (1986) p. 1047.
27. J. Stice, in *Engineering Aspects of Shape Memory Alloys*, ed. T. W. Duerig *et al.*, (Butterworth–Heinemann Ltd., 1990) p. 483.
28. *Proc. International Conference on Shape Memory and Superelasticity Technologies*, (Pacific Grove, California, 1994), in press.
29. J. Haasters, G. v. Salis-Solio and G. Bensmann, in Ref. [9], (1990) p. 426.
30. Shibi Lu, ibid., (1990) p. 445.
31. R. Sachdeva and S. Miyazaki, ibid., (1990) p. 452.
32. Chu Youyi, in Ref. [7], (1989) p. 569.
33. Y. Oshida and S. Miyazaki, *Corrosion Engineering*, **40** (1991) 1009.
34. K. M. Speck and A. C. Fraker, *J. Dent. Res.*, **59** (1980) 1590.
35. H. Kimura and T. Sohmura, *J. Japanese Society for Dental Materials and Devices*, **6** (1987) 73.
36. N. Fujita, Y. Sato, F. Uratani and M. Miyagi, *Report of Osaka Prefectural Industrial Technology Research Institute*, No. 86 (1985) 32.
37. J. Takahashi, M. Okazaki, H. Kimura, N. Horasawa, M. Ito and S. Takahashi, *J. Japanese Society for Dental Materials and Devices*, **5** (1986) 705.
38. K. Okano, T. Fukubayashi, S. Tomobe, T. Tateishi, Y. Shirasaki, K. Saito and M. Fukui, *Biomaterials*, **4** (1986) 113.
39. H. Yoshizawa, E. Shigeura, K. Suzuki, S. Fukuyo, K. Hashimoto and E. Sirenji, *Dental Implant*, **10** (1985) 12.
40. G. Bensmann and G. v. Salis-Soglio, *Tech. Mitt. Krupp. Forsch.*, **42** (1984) 25.

Index

282

d in the United States
okmasters